Triple Threat Power Grid Exercise

# High-Impact Threats Workshop and Tabletop Exercises Examining Extreme Space Weather, EMP and Cyber Attacks

## By InfraGard National EMP SIG TTX Planning Team

Dr. George Baker III; Terry Donat, MD; David Hunt; William Kaewert; Mary Lasky; Cedrick Leighton; Charles Manto; Dana C. Reynolds; Robert Rutledge

Edited by Feinman, Lasky, and Manto

Cover and title page photos © used by permission by John Coyle, coylestudios.com

Book cover and interior design by Jeffrey Barnes jbarnesbook.design

**Westphalia Press**
An imprint of Policy Studies Organization
1527 New Hampshire Ave., NW
Washington, DC. 20036
info@ipsonet.org

Westphalia Press
An imprint of Policy Studies Organization
1527 New Hampshire Ave., NW
Washington, D.C. 20036
info@ipsonet.org

ISBN-13: 978-1-63391-249-6
ISBN-10: 1633912493

Cover design by Jeffrey Barnes:
jbarnesdesign.com

Daniel Gutierrez-Sandoval, Executive Director
PSO and Westphalia Press

Updated material and comments on this edition
can be found at the Westphalia Press website:
www.westphaliapress.org

# High Impact Threats to the Electrical Grid

## Workshop & Tabletop Exercise

"The tendency in our planning is to confuse the unfamiliar with the improbable. The contingency we have not considered seriously looks strange; what looks strange is thought improbable; what is improbable need not be considered seriously."

~ Thomas Schelling, Nobel Prize Winner in Economics

Foreword to *Pearl Harbor: Warning and Decisions* by Roberta Wohlsteller
(Stanford: Stanford University Press, 1962)

## CALL FOR COLLABORATION

The InfraGard EMP SIG is providing these materials and
seeks to continually improve them.

Please send requests for the "Facilitator's Guide" and any proposed updates, lessons
learned, and insights gleaned from conducting the workshop and/or exercise to the
InfraGard EMP SIG at:
**igempsig@infragardmembers.org**

See "Conclusion and Survey" on page 109.

*Thank you!*

# Table of Contents

# Proposed Agendas

## Alternative 1—Workshop and Tabletop Exercise

**Registration, Check-In**

**Welcoming Remarks**

**Workshop Objectives**

**Workshop**

**Tabletop Exercise—Three Different Exercises Scenarios (Cyber, Solar, or EMP)**

**Tabletop Hot Wash**

**Closing Remarks and Adjourn**

## Alternative 2—Tabletop Exercise Only

**Registration, Check-In**

**Welcoming Remarks**

**Objectives and Background Information**

**Tabletop Exercise—Three Exercise Scenarios (Cyber, Solar, or EMP)**

**Tabletop Hot Wash**

**Closing Remarks and Adjourn**

# Acknowledgments

The InfraGard Electromagnetic Pulse Special Interest Group (EMP SIG) planners, practitioners, and subject-matter experts contributed to the design, delivery, and evaluation of this exercise program. We are excited by the prospect that this exercise material can and will be used around the nation, whether by local InfraGard Chapters, public and private-sector entities, local communities, or other advocacy groups concerned about high-impact threats to the electrical grid—our most vital, life-sustaining infrastructure. We look forward to compiling best practices, lessons learned, and bibliographical items for the bibliography from users of this material so that these resources may be updated and improved over time. Special thanks goes to Catherine Feinman, editor of DomesticPreparedness.com and the *DomPrep Journal*, for editing the materials.

Planning Team Members (see Appendix B for biographical information):

Dr. George Baker III
Professor Emeritus, James Madison University
Leader, EMP Technology, EMP SIG

Dana C. Reynolds
Director, Colorado Information
    Analysis Center
Administrative Officer, EMP SIG

Terry Donat, MD
Facial Plastic & Reconstructive Surgeon
Metropolitan Chicago, Illinois
Co-Chair, Health Industry Adv. Panel, EMP SIG

Cedrick Leighton
Chairman, Cedric Leighton International Strategies
Cyber Panel, EMP SIG

David Hunt
Homeland Security Consulting
Waterford, VA

Chuck Manto
CEO, Access Networks LLC
Chairman, EMP SIG

Bill Kaewert
President, Stored Energy Systems LLC
Strategic Officer, EMP SIG

Bob Rutledge
Lead, Space Weather Forecasting Office, NOAA

Mary Lasky
Program Manager, Johns Hopkins
Applied Physics Laboratory
Liaison Panel Facilitator, EMP SIG

# Introduction

In July 2011, the InfraGard National Board and the Federal Bureau of Investigation approved the formation of the InfraGard National Electromagnetic Pulse Special Interest Group (EMP SIG) for the purpose of sharing information about catastrophic threats to the nation's critical infrastructure and encouraging local communities to become more resilient. The threats specifically include manmade EMP, cyberattacks, coordinated physical attacks, pandemics, and extreme space weather.

This EMP SIG exercise concept is intended to provide participants with a forum to begin discussions, planning, and preparation for catastrophic events involving the electrical grid and the cascading impacts to other critical infrastructure and the community. This exercise concept includes three separate scenarios, which can be conducted independently or in any combination to examine how different causes of grid failure can affect local communities and warrant different preparedness efforts. The exercise can be tailored to the needs of the exercise participants, to build awareness and foster preparedness. Each scenario has been reviewed for accuracy by leading subject-matter experts in order to facilitate realistic discussion and dialogue on the characteristics and impacts of these events. An extensive bibliography has also been included to provide participants the opportunity to learn more about these threats, either before participating in the exercise or as a resource in the preparedness and planning process.

High-impact threats are categorically different from many other threats or hazards for one principal reason. Unlike Hurricane Katrina or Superstorm Sandy, which affected regions and allowed other areas of the nation to rally to the aid of local communities, high-impact events have the capacity to affect much of the country simultaneously, thereby limiting resources for response and recovery efforts. Instead of waiting days for help, affected regions could wait months for any meaningful aid. In a "just-in-time" society, the consequences are barely imaginable, but some historical background may help planners appreciate the need to minimize these effects.

In October 2011, the National Defense University and the EMP SIG co-sponsored the first comprehensive nationwide contingency planning workshops and exercise on extreme space weather that could have a nationwide impact. Until that time, even the U.S. Department of Defense had not planned for a collapse of civilian infrastructure nationwide that would last more than a couple of weeks (outside of nuclear or world war). In December 2011, the EMP SIG reported its findings in a seminar at the December 2011 Dupont Summit, hosted by the Policy Studies Organization in Washington, D.C. Less than eight months after the summit, on July 23, 2012, Earth experienced a near miss of a potentially cataclysmic solar storm. By the second anniversary of the solar near miss, an array of scientific articles provoked attention in the international media.

A NASA article published on July 23, 2014 quoted one of the *Space Weather* authors, Daniel Baker from the Laboratory for Atmospheric and Space Physics at the University of Colorado Boulder, "I have come away from our recent studies more convinced than ever that Earth and its inhabitants were incredibly fortunate that the 2012 eruption happened when it did. …If the eruption had occurred only one week earlier, Earth would have been in the line of fire." The NASA article cited the often-quoted 2008 National Academy of Sciences report on a Federal Emergency Management Agency (FEMA)-funded economic impact assessment, which stated that the total economic impact of such an event could exceed $2 trillion or 20 times greater than the cost of Hurricane Katrina. Multi-ton high-voltage transformers damaged by such a storm might take years to repair. Baker said, "In my view, the July 2012 storm was in all respects at least as strong as the 1859 Carrington event. … The only difference is it missed."

In another July 2014 research article published in *Space Weather*, entitled "Assessing the Impact of Space Weather on the Electric Power Grid Based on Insurance Claims for Industrial Electrical Equipment," the authors showed how even small space weather events have been causing damage to the electric power grids. Claim statistics from an examination of over 11,000 insurance claims from 2000 to 2010 revealed that "geomagnetic variability can cause malfunctions and failures in electrical and electronic devices that, in turn, lead to an estimated 500 claims per year within North America." If small events can have such an effect, it becomes a lot easier to imagine the impact of the storm that just missed Earth in 2012. In addition, this data suggests that, if protection were to be provided for equipment against the larger threat, then money would be saved on a day-to-day basis for even the smaller events.

Awareness of this storm peaked when *The Washington Post* editorial board made its recommendation on August 9, 2014, "The world can and should do more to prepare, adapting satellite systems, toughening electric grids and, above all, ensuring that scientists have the tools they need to anticipate space weather…. For a variety of reasons—including the threat of severely inclement space weather—lawmakers must take a wider view."

In April 2015, the U.S. government's National Science and Technology Council (NSTC) responded with its draft "National Space Weather Strategy" outlining a coordinated federal response to severe space weather that could create long-term regional or national blackouts. Its second goal of "enhancing response and recovery capabilities" includes: "Develop and conduct exercises to improve and test federal, state, regional, local, and industry-related space weather response and recovery plans" (see http://www.dhs.gov/national-space-weather-strategy).

Manmade EMP poses even greater problems according to studies publicly released by the congressional EMP Commission between 2004 and 2008 and highlighted in the August 14, 2014, *Wall Street Journal* op-ed by R. James Woolsey and Peter Vincent Pry, both formerly with the Central Intelligence Agency. Not only is it possible for small mobile electromagnetic interference devices to be used at relatively close range against vulnerable electronic equipment and systems, but a relatively small-yield nuclear weapon could be placed on a scud missile, launched from an offshore freighter, and detonated in the upper atmosphere (80–300 miles high) to impact multiple regions or an entire continent. The electromagnetic fields emanating from EMP weapons include those that are in the billionths of seconds—much faster than lightning strikes. They travel through the air and across any kind of conductor, particularly long power or communication wires that act as giant antennae to transmit the pulse to connected equipment.

A September 10, 2007 economic impact assessment by the Sage Policy Group of Baltimore, Maryland, showed that even a regional EMP incident between Richmond, Virginia, and Baltimore could cause $770 billion of economic damage, even without considering loss of equipment or secondary effects such as lack of water in a large fire. The EMP Commission gave high marks for the study methodology and results, as did the economists who did the work quoted by the Academy of Sciences. In addition, the Sage report determined that protecting even 10% of the most critical infrastructure could alleviate up to 85% of the economic losses in medium-impact scenarios. This Sage Policy Group study shows that it can be relatively inexpensive to protect critical infrastructure and that not all infrastructure may need to be protected to the same degree. However, as in the case of extreme space weather, little has been done until now to protect civilian critical infrastructure. Numerous studies have shown that U.S. lifeline infrastructures are highly interdependent and erected much like a "house of cards." Subsequent tests by Iran of freighter-launched missiles, North Korean satellite success, and turbulence in places such as the Middle East have increased concerns about the ability of nonstate actors and the likelihood of a high-altitude nuclear EMP event.

In the same way that the NSTC warned of long-term blackouts due to severe space weather, on May 26, 2015, the Defense Threat Reduction Agency (DTRA) warned of long-term blackouts due to EMP in their Small Business Innovation Research request for proposals on the topic DTRA152-006, "Island-mode Enhancement Strategies and Methodologies for Defense Critical Infrastructure." Their public overview included the following statement,

"An electromagnetic (EM) attack (nuclear electromagnetic pulse [EMP] or non-nuclear EMP [e.g., high-power microwave, HPM]) has the potential to degrade or shut down portions of the electric power grid important to the DoD. Restoring the commercial grid from the still functioning regions may not be possible or could take weeks or months."

Cyberattacks have affected everyone, but because their effects are often unnoticed, they are considered by many to be merely an inconvenience. That is a mistake based on a comprehensive misunderstanding of the true nature of the cyberthreat. In both 2013 and 2014, the Director of National Intelligence classified cyberattacks as being the #1 threat to the United States. The Director of National Intelligence considered cyberattacks to be a greater threat than a terrorist attack.

On November 20, 2014, Admiral Michael Rogers, head of the National Security Agency (NSA) and Commander of the U.S. Cyber Command, warned that China, along with other countries, have the capability to successfully launch a cyberattack that could shut down the electric grid in parts of the United States. Rogers indicated if the United States remains on the defensive, it would be a "losing strategy." At a House hearing, Rogers indicated U.S. adversaries are performing electronic "reconnaissance," on a regular basis so that they can be in a position to attack the industrial control systems that run everything from chemical facilities to water treatment plants. This characterization of the cyberthreat may not be shared by the general public or by U.S. corporate leaders. Although consumers have been largely spared the true costs of cybercrime and cyberattacks, their cost to the global economy could in fact be as high as $2 trillion per year, according to Kerridge Commercial Systems (KCS), a British provider of commercial intelligence products. The U.S. share of this may be as high as $1 trillion per year. Estimates of the cost of cybercrime vary widely because this type of crime is under-reported. Some businesses have had to cease operations because they became victims of cyberattacks. The case of Canadian telecommunication equipment manufacturer Nortel shows what can happen to a business if the cyberthreat is ignored. The company's leadership ignored a cyberthreat resident on their networks for over a decade. The result was the theft of billions of dollars' worth of intellectual property and the eventual bankruptcy of the company. That threat was traced to China, and now the Chinese telecommunication equipment manufacturer Huawei has a dominant position in some global markets that used to belong to Nortel. Insurance companies are only now beginning to grapple with the problem of providing adequate insurance coverage for cyberattacks.

Although there has been some success in this area, most insurance companies do not understand the true nature of the cyberthreat and have not yet established proper pricing mechanisms for cyberinsurance products. The largest risks to society are likely to be experienced in the arena of industrial control systems, which are largely unprotected by cyberthreat-prevention techniques. Numerous reports have shown that foreign cyberattackers have already breached many utilities. In early November 2014, ABC News and other media outlets reported on a hack of utility systems dubbed "Black Energy." According to the reports, the hack was traced to government-sponsored attackers from Russia. The hackers purportedly gained access several years ago to the Industrial Control Systems that run electrical utilities, oil pipelines, and refineries. Once they got into these networks, the hackers conducted what amounted to cyber reconnaissance missions, so they could diagram the various targeted computer networks. From there, they could find the networks' vulnerabilities.

The U.S. government has been aware of potential Industrial Control Systems network vulnerabilities for some time. A Federal Energy Regulatory Commission report, released on March 12, 2014, showed how a successful attack of only nine electric grid facilities could result in a nationwide power outage. The report published in *The Wall Street Journal* resulted in a hastily convened U.S. Senate hearing. There was no challenge to the accuracy of the report about the grave vulnerability the country faces, but rather only a challenge because the report was "mishandled" and leaked to the public. Although at the moment the vast majority of cyberattacks are low-impact, high-frequency events, there is a growing concern about them becoming high-impact, high-frequency events. Like other high-impact threats, they have the ability to cause similar levels of disaster, especially when combined with other threats. However, appropriate technical and organizational mitigation and thorough preparation of leadership and key personnel in the affected industries can reduce both the impact and the temptation for adversaries to try to attack them.

What remains uncertain is the willingness to engage these high-level threats. Psychological and political views complicate the discussion—a way to impose more government regulation versus a scare tactic to raise the nation's defense and homeland security budgets. These divergent political views help explain why no meaningful cybersecurity legislation has been signed into law in the last few years. In reality, there are daily cost savings, economic development benefits, as well as environmental and security benefits when taking a reasonable systems approach to mitigate these threats. This is especially true when local communities are more sustainable and capable of creating and managing a larger percentage of their critical power and food requirements. Similar to concerns that Senators have raised at past cyberthreat hearings, some think it is a challenge to begin an EMP discussion without causing panic or providing too much information to "the bad guys."

*This exercise concept may engage emergency management, first responders, infrastructure owners, political leaders, and other community leaders, who are already accustomed to disaster and continuity planning. It also could be used by those not experienced in disaster planning to help them think about the disaster and then start their own planning. This exercise serves as a first step to share factual information about the nation's vulnerabilities to these events and begin a planning process to improve preparedness within communities. The InfraGard EMP SIG will continue other efforts to engage these communities, through forums such as the Dupont Summit and other locally available forums.*

# Introduction

The *High-Impact Threats to the Electrical Grid Workshop and Tabletop Exercise* is designed to establish a learning environment for players to exercise emergency response plans, policies, or procedures and bring awareness of three potential catastrophic scenarios that could adversely impact the electrical grid, resulting in a regional or all-of-nation blackout scenario, impacting millions.

# Purpose

The purpose of this workshop and tabletop exercise or tabletop exercise alone is to provide a forum to both educate players and discuss various protection, mitigation, response, and recovery strategies related to a long duration (>30 days) electrical blackout stemming from either a geomagnetic storm, a cyberattack, or an EMP event. The exercise also will highlight the range of anticipated and unanticipated cascading effects that could occur during a long-term grid outage.

# Running the Exercise

Exercise leaders may choose to either run the exercise themselves, or request facilitation help from a qualified EMP SIG member. Depending on the exercise location and other factors, there may be charges for this assistance. If you are interested, please send email to: **igempsig@infragardmembers.org**

The following objectives apply to both the Educational Workshop and the Tabletop Exercise.

## GENERAL EXERCISE OBJECTIVES

- Develop an exercise toolkit (this document) on the GMD/EMP/cyberthreat scenarios that can be utilized by partner agencies, community groups, and others to conduct local or regional exercises on the topic of GMD/EMP/cyberthreat to inform future protection, mitigation, response, and recovery activities.
- Provide credible, read-ahead material from subject-matter experts on the GMD/EMP/cyberthreat.
- Conduct one or more of the exercise scenarios that will thoughtfully engage exercise participants and observers.
- Create a community of interest around the GMD/EMP/cyberthreat scenarios to stimulate unified action.

## CORE EXERCISE OBJECTIVES

- Stimulate public and private sector awareness of GMD/EMP/cyberthreat by effectively distinguishing between the threats and highlighting differential impacts depending upon event duration and magnitude.
- Identify practical improvements for the "whole of community" that would boost readiness and shorten recovery time.
- Identify how a prolonged supply chain disruption following a GMD/EMP/cyberthreat event would impact the responder community, infrastructure providers, and the nation.
- Identify and discuss principal cascading effects from a prolonged grid outage following a GMD/EMP/cyberthreat event. (Participants should focus on grid impacts and cascading effects on their organization.)
- Discuss whether the private and public sectors could facilitate sheltering and mass care during a prolonged grid outage.
- Have material on actions people can take including alternative energy sources and their efficacy during a prolonged grid outage. (This might include local or distributed energy sources. Microgrids may be integrated into or run separate from larger grids that may make use of wind, solar, passive solar, geothermal, waste-to-energy systems, tidal, and various kinds of energy storage systems that could be used for specific uses, buildings, campuses, neighborhoods, or towns.)

# Exercise Scenario Background Information/Read Ahead

## Cyberattack—SCADA Systems

Supervisory Control and Data Acquisition (SCADA) systems are real-time industrial process control systems used to centrally monitor and control remote or local industrial equipment such as generators, valves, pumps, relays, circuit breakers, etc. SCADA is used to control chemical plant processes, oil and gas pipelines, electrical generation, transmission and distribution equipment, manufacturing facilities, water purification and distribution infrastructure, etc. SCADA originally was designed for managing a single location's processes, but the cost and managerial benefits soon moved SCADA systems to monitor and control offsite, remote facilities. Industrial plant-scale SCADA is often referred to as a distributed control system or DCS.

Because SCADA systems were originally developed by engineers looking to improve process management, they were not developed using conventional computer programming formats. At the time SCADA systems were developed, they were for internal control, and computer security was not a major concern. That meant that even common computer security safeguards, such as the installation of normal firewalls, were not applied to SCADA systems. Because many of the SCADA devices have a life span of decades, there are millions of devices still in service that did not have security protocols built in, nor do they have sufficient memory to enable retrofits. Typical virus protection does not work with SCADA systems, and older SCADA devices do not have their passwords or their data encrypted. Many of the devices are sealed units with no capacity for upgrades. Some of the SCADA sensors or controllers are no longer supported by the manufacturers because they are considered to be outdated equipment, so retrofitting security protocols require a significant investment to replace operational equipment before its lifecycle expires. This requires a significant cost–benefit analysis, with perceived risk as a major variable. (*Note:* New standards are under development for complex distributed energy systems that are part of smart-grid systems that are beginning to address these issues. Future versions of these materials will take them into consideration as they become adopted.)

As the level of awareness of the vulnerability of some of these systems has grown, there has been a push to bring the SCADA systems under the umbrella of corporate IT security systems. However, the integration of SCADA systems into traditional IT security can create more problems than it solves. Traditional IT security methods and practices are not compatible when managing thousands of separate devices that need to be accessed continually and, in many cases, cannot tolerate down time. Normally, each device would have a unique password for access, but it is not practical for each operator to log into thousands of devices separately. In many cases, the devices cannot accept multiple passwords for authentication. These situations lead to problems when a single password accesses multiple devices, and the difficulty in changing passwords can permit former employees having current passwords to the systems.

In SCADA systems, constant access to the data feeds is required, as well as sending control signals to the remote equipment. For reliability, SCADA response to commands must be instantaneous. Because SCADA devices may monitor systems many times per second, any degradation to the signal quality between the monitoring station and the SCADA device or connection delays (such as navigating security protocols) may cause completion of an operation after a deadline, which can cause process disruption or severe cascading system failures.

The use of firewalls on these systems can create problems or delays in accessing remote systems, which has added a layer of unreliability to the systems. Some authentication systems have proved to be too time consuming and created delays in controlling systems, which is unacceptable. The enhanced use of biometric authentication and modern encryption techniques should become a normal component of SCADA systems. At some point, federal regulations may require such protections be built into SCADA systems.

In some cases, the IT staff has created animosity by applying normal IT solutions that complicated life for the engineers who have to run the systems. The normal practice of shutting down IT systems for weekly maintenance is not compatible with 24/7 manufacturing operations. This can lead to an "us-versus-them" culture, which is not productive for security of the systems.

Although the transition to traditional program languages has improved the ability to layer security measures on the monitoring and control systems, it also has increased the ability of hackers to use conventional methods to access or control the SCADA systems. Previously, "bedroom hackers" were not able or sufficiently interested in penetrating SCADA systems, so it took a more sophisticated and determined effort to understand the different programming used by SCADA installations. Now, as the SCADA systems are brought under normal IT security systems, accessing the corporate IT can access the SCADA data or controllers, using readily available hacking software programs. One method to counter this is to place "honeypot" or decoy systems on the network in order to monitor intrusions into the system.

Often the focus of corporate IT has been on protection of the main servers and not on remote locations, or "edge" client devices. However, in SCADA systems, the edge devices are often on the "front line" in managing the production. The energy sector saw an increase in cyberattacks on gas and petroleum pipeline systems starting in 2012. In Saudi Arabia and Qatar, two major oil and gas companies were victims of a cyberattack that featured the insertion of the so-called "Shamoon" virus via an infected USB drive. Saudi Aramco, the world's largest petroleum company, lost 30,000 PCs to the Shamoon virus, which reportedly originated in Iran.

A primary concern is that other nation states are well aware of the vulnerability of these systems to cyberintrusion, and have dedicated significant resources to be able to penetrate critical infrastructure controls. The Iranians, the Russians, and the Chinese are all reportedly targeting U.S. electrical utilities. All of these nations possess highly regarded offensive cyber capabilities. Future wars likely will be fought on many fronts including cyber.

The effects of a cyberattack can range from minor to catastrophic. A major, coordinated attack against the electric grid operators and the electricity producers could have an effect similar to that of a major solar storm in disruption of the grid and damage to the infrastructure. These outages cascade into virtually all other critical infrastructure sectors, which rely on electrical power. Standards for the integration of complex distributed energy elements and systems are underway within various industry groups. Working through these scenarios in workshops and exercises can be helpful for those in industry developing these standards and those dependent on these systems undergoing radical change.

## Geomagnetic Storms

In general, geomagnetic storms are disturbances in the Earth's normally dormant geomagnetic field caused by intense solar activity. A rapidly changing geomagnetic field over large regions will induce geomagnetically induced currents (GIC), a quasi-direct current to flow through interconnected electric power grids. GIC flow can cause widespread catastrophic damage to key power grid transformers causing restoration problems. A severe geomagnetic storm has the potential to be a worldwide event.

Space weather is the variable condition on the sun and in the space environment that can influence the performance and reliability of space-borne and ground-based technological systems, as well as endanger life or health.

Most space weather occurs because emissions from the sun influence the space environment around Earth, as well as neighboring planets. Space weather is not a new phenomenon—the impacts of these storms have been recorded in North America as far back as 1847. Nevertheless, space weather is a relatively new field in emergency management, which is rapidly being educated as to its significance.

Space weather includes solar flares, solar radiation storms, and geomagnetic storms. Extreme geomagnetic storms are low-frequency, high-impact incidents that have occurred with relative consistency throughout history. The word "geomagnetic" refers to the magnetic principles of the Earth, which are controlled by the rotation of the planet and the physical characteristics of the Earth's metal core. The Earth's magnetic field experiences normal fluctuations but, during a geomagnetic storm, the disturbances caused by space weather are so great that they generate electrical currents in the ground (known as GIC). These currents can disrupt or damage electrical grid components. Geomagnetic storms can affect technological systems based in space (e.g., satellites) and on the ground (e.g., power grids and communication lines) by interrupting their normal electronic and magnetic components.

Geomagnetic storm watches, warnings, and alerts are delivered via email to National Oceanic and Atmospheric Agency (NOAA) space weather product subscription service subscribers when geomagnetic storm levels are occurring or are expected to occur. FEMA receives email notification when G3 (strong) levels are occurring or are expected to occur and redundant email and telephone notification when G4 (severe) and G5 (extreme) levels are occurring or are expected to occur. State, territorial, tribal, and local jurisdictions can sign up on the SWPC website to receive email alerts.

A geomagnetic storm takes anywhere from 16 to 18 hours up to several days to impact Earth, with the more significant storming generally driven by storms with the fastest arrivals. These storms typically last 6 to 24 hours but, during periods of very high solar activity geomagnetic storms, can persist for days. The G4 and G5 geomagnetic storm levels have potential to cause or exacerbate a major disaster or emergency, interfere with or seriously degrade FEMA's response and recovery capability by causing widespread voltage control problems and protective system problems. Some grid systems may experience complete collapse or blackouts, and transformers may experience damage during the strongest of G5 geomagnetic storms. It should be noted that there is some saturation in the G5 storm category and only a fraction of the largest G5 storm values are anticipated to cause significant problems. In these large G5 storms, some grid systems may experience complete collapse or blackouts, and transformers may experience damage. In addition, severe geomagnetic storms can create political, public, or media pressure and expectation for government action. Critical infrastructure impacted by geomagnetic storms is as follows:

## Electric Power Grid

As mentioned above, widespread voltage control problems and protective system problems can occur, and transformers may experience significant damage. Large-scale geomagnetic storms can result in complete power grid collapse or blackouts. In addition, pipelines and railways also experience effects from strong geomagnetic storms. Regions in the Northeast United States from the Chesapeake Bay area through the Great Lakes states are the most vulnerable. Northwestern states also are vulnerable. This is the case for the following reasons:

- high latitudes have greater vulnerability to geomagnetic storm disturbances;
- close proximity to ocean salt water as a large conductor of electricity;
- geology with rock formations that are more conductive to the flow of GIC; and
- extensive interconnectivity of the grid in the Northeast United States, which can result in powerful cascading effects.

## GPS Applications

GPS can be significantly degraded, especially in the polar and auroral regions, which can extend to the U.S./Canadian border during severe geomagnetic storms. Single-frequency GPS receivers are particularly vulnerable because unlike the low-frequency radio transmissions used by terrestrial systems, GPS uses radio signals that pass through the ionosphere. Geomagnetic activity can affect the character of the ionosphere and, consequently, the proper function of navigation systems. Certain storms may result in "loss of lock and no position information," even at lower latitudes across the United States.

## Satellite Communications

Short-lived periods of degraded Satcom are possible. Services reliant on both Lower Earth Orbiting (LEO) satellite constellations and satellites in Geostationary (GEO) orbit may be affected. Satellite electronics may be affected resulting in damage, or even loss of the satellite.

## High-Frequency (HF) Communications

May become unreliable, or even totally unusable, due to impacts on the electrical characteristics of the ionosphere. Frequent solar flaring during periods of elevated solar activity often renders HF unusable on the sunlit side of Earth for tens of minutes to several hours at a time. The solar radiation storms that often occur in conjunction with significant activity also can render HF unusable at high latitudes (effect strongest from 60° to 65° latitude and poleward). Geomagnetic storms can also affect HF propagation on both the day and night side of Earth and can have unpredictable impacts at essentially all latitudes.

## Aurora

May be visible at mid-latitudes including southern-most states such as Florida and Texas.

# Societal Impacts of Extreme Space Weather Incidents

Historical records indicate that space weather incidents throughout history have been considerably more intense than anything seen in the last 50 years. Recent research suggests that if a storm like the 1859 or the 1921 storms occurred today, the consequences to a technology-dependent society would be significant. The effects of extreme space weather are best classified in the categories of "primary" and "cascading" or secondary effects.

Primary effects are those that are a direct result of space weather (e.g., radio blackouts, electric power-grid disruptions, GPS disruptions, etc.). Cascading effects are those that result from the primary effects (e.g., business, government, and hospitals not being able to function properly, transportation disruptions, etc.). Much of the below information is adapted from a National Academy of Sciences workshop report titled, "Severe Space Weather Events—Understanding Societal and Economic Impacts." For a PDF of this report, go to: **http://www.nap.edu/catalog/12507.html**

Extreme space weather events are low-frequency/high-consequence incidents that present a unique set of problems for public and private institutions and government agencies. Because systems can quickly become dependent on new technologies, vulnerabilities in one part of the broader system have a tendency to spread to other parts of the system. The societal and economic impacts of space weather depend on a wide range of factors such as:

- the magnitude, duration, and timing of the incident;
- the nature and severity of the effects;
- the robustness and resilience of the affected infrastructures;
- the risk management strategies and policies that the public and private sectors have in place; and
- the capability of the responsible federal, state, territorial, tribal, and local government agencies to respond to the effects of an extreme space weather event.

Because of the interdependencies of critical infrastructures in modern society, the impacts of severe space weather events can result in short-term as well as long-term socioeconomic disruptions. For example, much of modern society relies on electricity to function. Even though it is unlikely that extreme space weather could result in a large-scale electric blackout, the consequences of this kind of incident would be devastating and would cascade through other dependent systems. The effects of a long-term outage would likely include:

- disruption of the transportation, communication, banking, and finance systems;
- disruption of government services;
- the breakdown of the distribution of potable water and sewage treatment systems due to pump failure; and
- food and medication shortages.

The loss of these services for a significant period of time in one region of the country having a cascade or ripple effect throughout other areas of the nation; extending ultimately into and across international infrastructures.

## Primary Effects

### Energy System Vulnerability

Energy providers place a strong emphasis on building robust and resilient systems, particularly for well-known and frequent threats (e.g., weather disruptions, human error, etc.). The threats posed by space weather, however, differ from those associated with terrestrial weather, particularly in the case of geomagnetic storms. These storms can develop quickly, with wide-reaching effects across entire regions and even entire continents. Energy systems are generally not designed to withstand simultaneous loss of many key assets, and this kind of impact potentially could result in widespread power disturbances and outages.

Only a limited amount of electricity is stored in the U.S. grid. Electricity availability depends on conversion from other energy sources (e.g., hydro, fossil fuel, nuclear) and the production of electrical energy must be instantaneously matched to the current demand. As it transports via the electric power grids of the United States and Canada, it requires constant attention within power stations to the many details associated with its transfer to assure safe, reliable, secure operations. For example, the demand for electricity in North America has grown dramatically over the last 50 years. The extra high voltage (EHV) infrastructure has grown to support these demands and, in a similar manner, as the grid has increased in size, it also has grown more coupled and vulnerable to disruption due to geomagnetic disturbances. As the bulk power system has grown in size, it has also grown in complexity, which has further compounded risks from geomagnetic disturbances. Some of the more important system changes that can increase impacts from geomagnetic events include higher design voltages and the behavior of transformers as voltage ratings increase. After the March 1989 solar storm, the electric sector developed procedures to operate the system in a conservative state with sufficient advanced notice of a geomagnetic disturbance. These plans have been largely effective at avoiding widespread blackouts to the system during the smaller and lower intensity geomagnetic storms that have occurred since 1989. However, these procedures were not designed for extreme levels of disturbance.

Contemporary models of large power grids and the electromagnetic coupling to these infrastructures by the geomagnetic disturbance (GMD) environment have matured to a level in which it is possible to achieve accurate benchmarking of geomagnetic storm observations. These efforts have allowed additional insights into the potential impacts to today's infrastructure that could result from large historically observed events.

Depending on the location and pattern of the geomagnetic disturbance, there are a number of plausible consequential outcomes for a severe geomagnetic storm of a strength roughly 10 times what was observed in 1989. A simulation based on the 1859 Carrington storm severity, shown in Figure 1, shows the pattern of GIC flows in the U.S. power grid and the boundaries of regions of power grid that could be subject to progressive collapse, such as what occurred to the Québec Interconnection in March 1989. The simulation results indicate that more than a thousand EHV transformers will have sufficient GIC levels to simultaneously be driven into saturation. Further, this would suddenly impose an increase of over 100,000 MVARs (megavolt-ampere reactive, a measure of electric power) of demand on the system, a scenario that could trigger a widespread voltage collapse, resulting in system instability and a blackout.

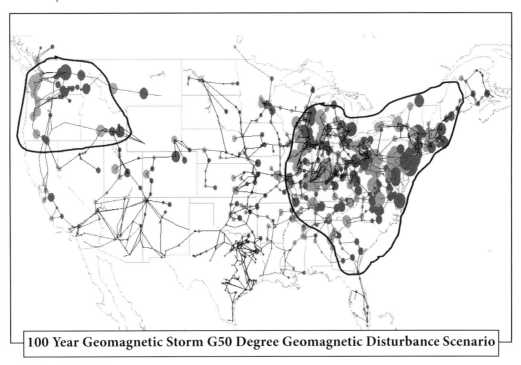

**100 Year Geomagnetic Storm G50 Degree Geomagnetic Disturbance Scenario**

**Figure 1:** The simulation results showing the pattern of GIC flows in the U.S. grid for a 4,800 nT/min geomagnetic field disturbance at 50o geomagnetic latitude, a storm approximately 10 times higher than the March 13, 1989 storm and representative of an 1859 class storm. The above regions outlined are susceptible to system collapse due to the effects of the GIC. *Source:* Courtesy of Metatech Corporation

Although the electric sector has performed reliably through all solar storms since the March 1989 event by using specialized geomagnetic storm operating procedures, all subsequent storms have been much lower in intensity than the March 1989 storm. The storms of concern that are addressed in this exercise could potentially be 4–10 times more intense than the March 1989 solar storm and entail the potential for widespread damage to EHV transformers and other key assets of unprecedented proportions. Permanent damage to these key assets could lead to power grid restoration problems and the possibility of very prolonged outages of electric power supply over large regions of the United States.

Both the 1989 and 2003 storms were small compared to the 1859 "Carrington Event." This superstorm was presaged by a dramatic increase in sunspot activity near the peak of a solar maximum. The sunspots were so large they could be seen with the naked eye. They spawned two consecutive coronal mass ejections (CMEs) heralded by immense solar flares observed by British astronomer Richard Carrington. When the first CME slammed into Earth several days later, the Aurora Borealis lit up from Maine to Florida. Magnetic recorders around the world shot off scale, and spurious electric currents overloaded the world's telegraph systems. When the second CME arrived six days later, it reinforced the first, turning night into day as far south as Panama. People could read the newspaper by the crimson green light. Telegraph systems became unusable across Europe and North America,[1] with many bursting into flames from the excess current.

---

[1] James L. Green, and Sten F. Odenwald, "Bracing the Satellite Infrastructure for a Solar Superstorm," Scientific American, August, 2008, p. 80-87.(July 28, (for Internet version).

During the period from 2012 to 2014 there were a large number of major solar flares of a Carrington scale, which missed the earth by only a few days. Because there are major solar storms on a regular basis, most experts believe it is likely that the Earth will experience extreme space weather events in the future.

## Satellite Disruption

Geomagnetic storms can change the dynamics of the trapped radiation environment around Earth, presenting one facet of increased risks to satellite systems as a result of space weather. Additionally, the solar radiation storms that almost always occur in conjunction with significant geomagnetic storms can cause a variety of system issues with impacts ranging from momentary interruptions to total satellite failure. These disruptions directly impact the designers and operators of satellites. Indirect impacts are described in the Communications Disruption section below.

## Signal Disruption

Space weather can affect HF radio communications at all latitudes. Television and commercial radio stations usually are not affected by solar activity, but ground-to-air, ship-to-shore, and amateur radio often are disrupted.[2]

## Electromagnetic Pulse

An EMP is a burst of electromagnetic radiation. This abrupt pulse of electromagnetic radiation usually results from certain types of high-energy explosions, such as a high-altitude nuclear explosion, or from a suddenly fluctuating magnetic field, as can be caused by a severe solar geomagnetic storm. These rapidly fluctuating electrical fields and magnetic fields may couple with electrical/electronic systems to produce damaging current and voltage surges to electrical or electronic equipment or systems. More specifically, an EMP may contain some or all of the following components.

## E1—"Fast Pulse"

The E1 "fast pulse" component is a brief and intense electromagnetic field that can quickly induce very high voltages in electrical conductors. The broadband nature of E1 (0–1 GHz) allows coupling to system conductors ranging from centimeters to kilometers threatening systems in general, not just long lines. The E1 component causes electrical breakdown voltages to be exceeded disrupting or destroying computers and communications equipment. E1 also can damage heavy-duty electrical equipment such as transformers, motors, and generators by initiating arcs that damage electrical winding insulation. Since system power supply current can follow E1-initiated arc paths (sometimes referred to as "power follow"), the energy contributing to system damage can exceed incident E1 energy. E1 changes are too fast for many ordinary protection systems to provide effective protection against it. Protection systems that are independently tested and rated for E1 should be used for critical systems expected to work during or survive an E1 event. E1 is associated with EMP caused either by a nuclear detonation (most effectively in the upper atmosphere) or by directed energy weapons (that may extend to 10 GHz or beyond) and is not associated with solar events.

## E2—"Intermediate Time"

The E2 component is an "intermediate time" pulse that can last from about 1 microsecond to 1 second after the beginning of the EMP. Because of the similarities to lightning-caused pulses and the widespread use of lightning protection technology, the E2 pulse is generally considered to be the easiest to protect against. The main potential of the E2 component is the fact that following a nuclear detonation-caused EMP, the E2 follows the E1 component, which may have damaged the devices that would normally protect against E2. E2 is also associated only with a nuclear detonation.

---

[2]  A Profile of Space Weather.  NOAA Space Weather Prediction Center.

## E3—"Slow Pulse"

The E3 "slow pulse" component is different from the other two major components found in a nuclear EMP. The E3 component of the pulse lasts tens to hundreds of seconds and results from distortion of the Earth's magnetic field by the nuclear fireball and heaving of ionized layers of the atmosphere. The E3 component amplitude and Earth coverage are similar to solar superstorm GMD effects. Both nuclear E3 and solar GMDs couple efficiently only to long conductors (greater than a few kilometers) and are a concern for long power lines, long-haul communication lines, and pipelines. They are known to directly damage transformers and long-haul communication line repeaters. They also can damage "uninterruptable" power systems (UPS) by causing reactive power instabilities in the electric power grid.

Note that radiation from high-altitude bursts is a threat to the operation of communication satellites, particularly those in Low Earth Orbit (LEO). Thus, reliance on satellite communications in lieu of EMP/GMD-debilitated, ground-based long-haul communications is problematic.

Similar E1 effects can be created by portable electromagnetic weapons; however, their range is much shorter. A coordinated physical attack could have the same effect as a nuclear EMP.

## Cascading Effects of Electrical Power Outages

Loss of electrical power, even in relatively small areas for just a few days, can have a profound effect on people and their communities. This impact is compounded in summer or winter by the loss of heating or air conditioning in homes, businesses, government facilities, etc. Serious threats to healthcare and supplies of food and water result from energy loss, and businesses cannot operate properly, particularly if transportation systems are incapable of moving people to and from work. In addition, the loss of telecommunications would have serious impacts and would hinder the emergency public information efforts of the government. Other examples of cascading effects include:

- food supply chain disruption;
- fuel supply chain disruption (e.g., loss of gas pumps, transporting of fuels, pipeline control disruption, etc.);
- transportation system disruption (roadway congestion, fuel depletion, electronic communication and controls);
- water systems offline;
- waste or sewage systems down;
- financial systems information management;
- refrigeration loss (a major critical infrastructure element);
- strained emergency services and potential loss of command and control;
- seriously challenged medical care after 72 hours;
- home dependency on electrical and rechargeable battery operated medical, perishable foods, lighting, communication, etc. devices challenged in the long term;
- aviation vulnerabilities;
- Global Positioning System (GPS) vulnerabilities;
- telecommunications and Internet vulnerabilities.

As the nation's infrastructures and services increase in complexity and interdependence over time, a major outage of any one infrastructure will have an increasingly widespread impact. For example, the dependence of nearly all critical services on information technology is ever increasing, and the flow of information is itself dependent on communications infrastructure and a reliable supply of electric power. Backup power supplies do exist, but in most cases only for limited periods. Service reliability includes provisioning of backup facilities, which must be sufficiently isolated from each other that a single and perhaps even multiple events would not simultaneously shut down both locations.

In addition to the above, space weather and EMP can create socioeconomic disruptions due to the interdependence of infrastructures. Banking and finance systems as well as government services would be impacted. Airline flights would be diverted from the affected region, and the loss of commercial communications satellites would affect business transactions. Manufacturing would be impossible with neither electricity nor computers and SCADA, and broadcast and print news media would be seriously hampered if not completely disabled.[3]

Loss of these systems for a significant period of time in even one region of the country could affect the entire nation and have international impacts. For example, financial institutions could be shut down, freight transportation stopped, and communications interrupted, as suggested in Figure 2. The concept of interdependency is evident (for example) in the unavailability of water due to long-term outage of electric power and the inability to restart an electric generator without water on-site, supplies of which have been exhausted.

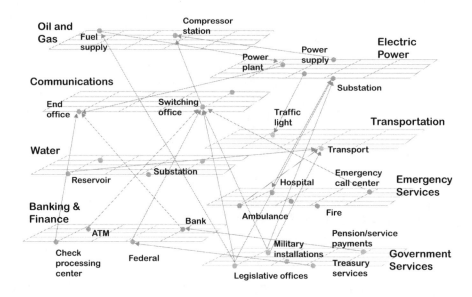

**Figure 2:** Connections and interdependencies across the economy. Schematic diagram showing selected interconnected infrastructures and their qualitative dependencies and interdependencies. *Source:* www.fcc.gov

## Communications Disruption

As a result of terrestrial and satellite communication network disruption, the following services can be hindered:

- communication services that provide news, education, and entertainment (e.g., global cellphones, satellite-to-home television and radio, and distance learning);
- a cost-effective means for interconnecting geographically distributed business offices (e.g., satellite links of store registers to regional distribution centers provide automatic inventory control and pricing feedback at a major retailer, and a major auto maker utilizes a satellite-based private communication network to update its entire system of dealer sales staff on new model features and service crews on new repair procedures);
- a cost-effective means of connecting businesses with their customers (e.g., facilitating point-of-sale retail purchases made with credit or debit cards at gas stations and convenience stores); and
- critical backup to terrestrial cable systems vital to restoring services during catastrophic events (earthquakes, hurricanes) that damage ground-based communications systems;[4]
- resynchronization of communication networks and electric power generation would be impaired if time-standard sites, ground-based communications connectivity, and GPS satellites are debilitated.

---

[3] "Severe Space Weather Events—Understanding Societal and Economic Impacts": for a PDF of this report go to http://www.nap.edu/catalog/12507.html.

[4] National Academy of Sciences, "Severe Space Weather Events—Understanding Societal and Economic Impacts," *Workshop Report,* 24, http://www.nap.edu/catalog/12507.html.

## Food and Water Shortages

With a massive electrical collapse, water distribution would fail, with obvious widespread public health consequences to the communities involved, including drinking water, sewage, and firefighting. Food supplies also would be depleted in rapid order. Without power, perishable and frozen food in homes, supermarkets, and warehouses that could not be consumed within days would rot. Canned and boxed food would be depleted next. A parallel collapse of logistical freight systems would seriously limit replenishment of sufficient food supplies.

## Navigation Systems

Systems such as Long Range Navigation (LORAN) can be negatively impacted by space weather, giving airplanes and ships information that is inaccurate by several miles. If navigators are alerted about radiation or geomagnetic storms, they can use a backup system to avoid any issues. In particular, geomagnetic storms can severely disrupt GPS applications such as the following:

- Clocks that keep the time to within 100 billionths of a second, without the cost of owning and operating atomic clocks. This capability is of enormous value to firms that need to synchronize their network computers or instruments, including the financial sector, for time-stamping electronic trading.
- Technology that has revolutionized transport logistics by making it possible to track and forecast the movement of freight.
- Precision agricultural methods of planning, field mapping, soil sampling, tractor guidance, crop scouting, and yield mapping.
- Maritime navigation, critical given the nation's reliance on imported oil carried by tankers and the environmental consequences of oil spills.[5]

## Healthcare Disruptions

The loss of hospital care, for example, is life threatening. Once emergency generators fail because fuel distribution for resupply has collapsed, life support systems, emergency room equipment, monitoring devices, and computers that keep hospitals operating would collapse. Surgery would become essentially impossible to perform. Many patients in medical facilities would die, and people with serious injuries or illnesses would have nowhere to go for definitive care. Disaster plans for hospitals undergoing a long-term power outage normally call for evacuation to other hospitals. But, if the other hospitals and other community locations are without power, also, then evacuation choices would be severely limited.

## Logistics Disruptions

Fuel supplies for power generation, individual transportation, and day-to-day logistical distribution systems would be interrupted. That, in turn, would cause a parallel disruption of every kind of freight distribution. Enterprises ranging from telecommunications to medical care that rely on backup generators would be unable to replenish vital fuel supplies. Due to the distance that many people in the United States have to drive to work, the disruption of individual gas/electrical transportation would in turn diminish workforce capacity in enterprises of every kind and, without electric power, most enterprises eventually would have to close. With a widespread failure of fuel distribution, the ability of freight carriers to replenish even critical supplies to communities would be seriously hampered. A long-term disruption of electrical power could spread from a single region of the country to affect the entire nation, with catastrophic consequences in public and private sectors.

## Socioeconomic Disruptions

The risks posed to society today by space weather and EMP are considerably different from those of the Carrington incident in 1859. Knowledge of the social, institutional, and policy implications of space weather is growing, but is far from being complete. Debates continue on how to address the issue, and who might have the best insights for detection, response, and mitigation.

---

[5] National Academy of Sciences. "Severe Space Weather Events—Understanding Societal and Economic Impacts," Workshop Report, 25. http://www.nap.edu/catalog/12507.html.

Disruptions to the telegraph system in 1859 caused communication issues. Because modern society is so dependent on large, complex, and interconnected technical systems—and because these systems not only are vital for the functioning of the economy, they also are vulnerable to electromagnetic events—a contemporary repetition of the Carrington event would cause significantly more extensive (possibly catastrophic) social and economic disruptions.[6]

## Learn more at:

**Geomagnetic (solar) Storms:** http://www.youtube.com/watch?v=s9YGOUWpH8s
**Electromagnetic Pulse:** http://www.c-span.org/video/?301046-1/electromagnetic-pulse-threat
**Cybersabotage:** http://www.c-span.org/video/?314419-1/electric-grid-cybersecurity-michael-hayden-industry-perspectives

## Recommended Exercise Read-Ahead Material

*This list is accessible, along with additional resources, at*
**http://www.ipsonet.org/conferences/the-dupont-summit**
*then click through to InfraGard segments at each of the summits from 2012.*
*See also the EMP SIG website at the secure InfraGard portal,* **www.infragard.org**

Electromagnetic Pulse Commission: Report of the Commission to Assess the Threat to the United States from Electromagnetic Pulse (EMP) Attack. Critical National Infrastructures, 2008. **http://www.empcommission.org/docs/A2473-EMP_Commission-7MB.pdf**

Electromagnetic Pulses—Six Common Misconceptions. Baker G. Domestic Preparedness, 2014.
**http://www.domesticpreparedness.com/Commentary/Viewpoint/Electromagnetic_Pulses_-_Six_Common_Misconceptions/**

Interdependence of the Electricity Generation System and the Natural Gas System and Implications for Energy Security. Judson N. MIT Lincoln Laboratory, 2013. **https://www.serdp-estcp.org/content/download/19069/.../TR-1173.pdf**

National Science and Technology Council; "National Space Weather Strategy", 2015. **http://www.dhs.gov/national-space-weather-strategy**

Physical Security of the U.S. Power Grid: High-Voltage Transformer Substations. Parformak P. Congressional Research Service, 2014. **http://fas.org/sgp/crs/homesec/R43604.pdf**

Severe Space Weather Events—Understanding Societal and Economic Impacts, NRC-NAP, 2008.
**http://books.nap.edu/catalog/12507/severe-space-weather-eventsunderstanding-societal-and-economic-impacts-a-workshop**

Solar and Space Physics: A Science for a Technological Society. NRC-NAP, 2013. **http://www.nap.edu/catalog/13060/solar-and-space-physics-a-science-for-a-technological-society**

---

[6] "Severe Space Weather Events—Understanding Societal and Economic Impacts," 29, for a PDF of this report, go to http://www.nap.edu/catalog/12507.html.

Solar Storm Near Miss and Threats to Lifeline Infrastructure. Manto C. Domestic Preparedness, 2014.
http://www.domesticpreparedness.com/Infrastructure/Cyber_%26_IT/Solar_Storm_Near_Miss_%26_
Threats_to_Lifeline_Infrastructure/

Solar Storm Risk to the North American Electrical Grid. Lloyds, 2013. http://www.lloyds.com/~/media/
Lloyds/Reports/Emerging%20Risk%20Reports/Solar%20Storm%20Risk%20to%20the%20North%20Ameri-
can%20Electric%20Grid.pdf

Space Weather Prediction Center, National Weather Service, NOAA. http://origin-www.swpc.noaa.gov/

Terrorism and the Electrical Power Delivery System. NRC-NAP, 2012. http://www.nap.edu/catalog/12050/ter-
rorism-and-the-electric-power-delivery-system

The EMP Threat: Examining the Consequences. Subcommittee on Cybersecurity, Infrastructure Protection
and Security Technologies of the Committee on Homeland Security, House of Representatives, 112th Congress, 2012. http://www.gpo.gov/fdsys/pkg/CHRG-112hhrg80856/html/CHRG-112hhrg80856.htm

The U.S. Government Thinks China Could Take Down the Power Grid.
http://www.cnn.com/2014/11/20/politics/nsa-china-power-grid/index.html?hpt=hp_t2

# Workshop

## Workshop Purpose

The purpose of this workshop is to provide a forum to learn about and discuss three high-impact threat scenarios, each with the potential to take down the electrical grid for an extended period (30 days or longer), negatively impacting the homeland on an unprecedented scale and triggering a series of cascading effects that would amplify impacts to the population.

## Workshop Objectives

The workshop has the following overall objectives:

- Examine the central differences and relative impacts of acts of cybersabotage, geomagnetic (solar) superstorms, and EMP events and their effect on the electrical grid.
- Discuss current levels of prevention, protection, and mitigation strategies as they apply to each of the three threat scenarios.
- Examine vulnerabilities in the electrical grid that make a long-term grid outage inevitable.
- Identify an ideal future state of preparedness that balances innovation, increased self-reliance, regulatory enhancements, operational procedures, hardening of the electrical grid, and recovery processes to improve grid resilience in the event of high-impact cyber, GMD, or EMP threats.

## Workshop Outcomes

The workshop focuses on the following outcomes:

- Educate participants on features of high-impact threats to the electrical grid caused by solar weather, EMP, and GMD.
- Enhance awareness of high-impact threats to the electrical grid, and their potential to trigger a long-term power outage.
- Identify effective and ineffective strategies to protect the delivery of electrical power from these threats/hazards.
- Determine the nation's readiness to respond and recover from any of these catastrophic threat/hazard scenarios.

## Running the Exercise

Exercise leaders may choose to either run the exercise themselves, or request facilitation help from a qualified EMP SIG member. Depending on the exercise location and other factors, there may be charges for this assistance. If you are interested, please send email to: **igempsig@infragardmembers.org**

*Note:* Local deliveries of this workshop may choose to use the subject-matter expert material from the national InfraGard EMP Special Interest Group, or may use presentations from local subject-matter experts, or a combination of presentations. Videos from the EMP SIG's Workshop on December 4, 2014 are available at: **https://share.dhs.gov/infragard-empsig-1204-052014/.**

# Tabletop Exercise Scenarios

Note: Local planners are encouraged to use one of the following three exercise scenarios: cyberintrusion, severe solar storm, or EMP attack. Planners may choose to use more than one of the scenarios but, due to the similarities of the three scenarios (massive outage affecting the population), it is recommended to conduct only one scenario per exercise, and to use the remaining scenarios for future exercises, building on what was learned during the workshop and tabletop discussions.

If the exercise participants are experienced in responding to catastrophic scenarios, any of the scenarios may be appropriate. If many of the participants are new to exercises, use of the solar or cyber modules is recommended, as the EMP incident can be overwhelming as a starting point.

Each of the following scenarios is divided into three modules, with a discussion period and proposed questions for each module. The first module covers the first week or two following the incident when the full impact is not yet experienced but beginning to be understood. The second module covers the following weeks through the first month or two. The final module covers from the end of a second or third month through the first year and recovery period (which would likely be years).

There are times when multiple disasters happen. Consequently, exercise planners may want to use the exercise coupled with a physical attack or natural disaster such as a severe winter storm or hurricane at the same time. This concept is more appropriate for experienced exercise participants, as combination scenarios can overwhelm some participants, resulting in a degraded learning environment.

# Cyberintrusion on the Electrical Sector Tabletop Exercise (TTX)

# Cyberintrusion in the Electrical Sector
## Module One

### Questions for Participants to Consider During This Module

1. What issues would you and your family face?

2. What would your organization do if the power were out for a week or two?

3. At the end of week one, what things might you be doing to prepare for the continued lack of power?

## Scenario Background

On April 8, 2009, the *Wall Street Journal* reported that cyber spies from several countries have penetrated the electrical grid and left behind software programs that could be used to disrupt the system. "They have attempted to map our infrastructure, such as the electrical grid," said a senior intelligence official. The espionage appeared pervasive across the United States and does not target a particular company or region, said a former Department of Homeland Security official. "There are intrusions, and they are growing," the former official said, referring to electrical systems. "There were a lot last year."

On October 29, 2014, the Department of Homeland Security announced that several industrial control systems—vendor-issued programs used by private companies to manage internal systems—had been infected by a variant of a Trojan horse malware program called BlackEnergy. Infected programs such as GE Cimplicity, Siemens WinCC, and Advantech/Broadwin WebAccess have been used by companies responsible for portions of the country's critical infrastructure. BlackEnergy is similar to a malware called Sandworm—which was used during a 2013 cyber espionage campaign against NATO, the European Union, and overseas telecommunication and energy sectors—that DHS believes could be "part of a broader campaign by the same threat actor." DHS said that although the Black Energy hacking campaign has been ongoing since 2011, no attempt appears to have been made to activate the malware to damage, modify, or otherwise disrupt industrial control processes.

A massive blackout struck Pakistan on January 25, 2015, leaving as much as 80% of the country without electricity. A militant attack on a transmission tower in the southwestern Baluchistan province caused the blackout, said Zafaryab Khan, a spokesman for the Ministry of Water and Power (*Source:* ABCnews.go.com, January 25, 2015). In the early morning hours today, many electrical operators across the United States notice erratic operation of their systems, and report this to NERC and the Energy Department.

The nation appears to be experiencing widespread targeting of electrical generation and transmission systems. Multiple utilities report malfunction of both computer controls and communication links essential to coordinated operation of the power grid.

Harried computer forensics experts determine that many utility-owned computers in key control centers are operating slowly, or have become unresponsive. Experts suspect malware, but there is neither obvious new intrusion nor distributed denial of service Internet attack.

During attempts to determine the cause of the problem, several large generating stations catch fire, which creates the belief that there may be a simultaneous physical attack on the system. Many large multi-megawatt generators are suffering phase synchronization problems with many damaged beyond repair.

As reports of system malfunctions pour into the regional and national centers, the nation experiences a near collapse of the eastern electrical grid. For unknown reasons, parts of the western grid continue to function, but there are significant outages in the Rocky Mountain States and in California, particularly in the Los Angeles and San Francisco Bay areas.

Many generating plants have been heavily damaged or destroyed. Computers and communications at major utility control centers are inoperable.

Power to some communication systems, water treatment facilities, hospitals, financial institutions, and emergency service agencies is supplied by on-site backup generators. On-site fuel is typically sufficient for 24–48 hours of operation before refueling is necessary.

## Impact Analysis—Day One

Much of the nation awakes to cold homes and low water pressure, but assumes the outage will be restored in a matter of hours. Those with backup power, smart phones or battery-operated radios learn the extent of the outage, that there may be difficulty in restoring power in the near future. Many people rush to the few stores that are open to buy food, propane, camping supplies, whatever they can with limited cash, or those stores that will accept checks or credit.

- Backup generators kick in and critical facilities are operational, with hours or days of fuel supplies.
- Most ATMs, gas stations, and stores are closed, without power or without staff.
- Traffic becomes snarled in cities without backup power for most intersection traffic signals.
- Critical workers attempt to get to their jobs, but are delayed due to the travel conditions.
- FEMA attempts to define the extent of the outage, and relays information to State Emergency Operations Centers across the nation.
- Much of the eastern half of the nation, plus Colorado and California are without power, with some small, rural systems remaining operational.
- The electrical industry is working to assess which specific issues caused the outage.
- The FBI, working with foreign intelligence agencies, is trying to identify the source of the hack, which clearly has been a coordinated, simultaneous attack on the nation's electrical grid. Fear that bringing the system back online without resolving the vulnerability could create further damage.
- Restoration crews are determining how they can patch the system together to begin a partial restoration of power. The extent of damage to the system remains unclear, as they try to assess which of the thousands of remote devices were damaged, and what remains in usable condition.
- Schools are closed; most businesses are closed as the nation begins to take stock of the situation. A feeling of being attacked, similar to 9/11, begins to take hold. Neighbors help each other with caring for the elderly and those with special needs.

## Summary—Day One

There are massive outages across the eastern United States, the Rocky Mountain States, and California. Many multi-megawatt generators are believed to be out of service. Parts of the country were less affected, including Texas, some parts of the west and rural areas served by smaller system generators. Slightly over 50% of the nation is without power. On-site diesel generators are expected to provide backup power to critical facilities for at least 72 hours, although most water utilities are down. Most hardwire phone services are operational and backup generators typically have fuel for 72 hours. Cell service is sporadic in the affected areas. Internet connectivity is also disrupted in some areas.

Mass transit in metropolitan areas has shut down. Weather is relatively warm for this time of year over much of the nation; thus, heating is not a critical concern at this time. System damage assessments are still underway to see what can be restored, and cyber experts in the industry and at national agencies are determining exactly what was done to cause the systemic damage.

# End of Week Two

## Communications
Most communications systems are running out of backup fuel supplies for generators. A few systems in areas with power are operational, but people are not able to complete many calls, due to telecommunication network failures. Lack of communications is hampering the recovery efforts. Government communications with the public are sporadic, with most television and radio stations without power in the affected areas.

## Care Facilities
Care facilities providing care to several million elderly and disabled have lost power to critical equipment and systems. Backup generators usually had only a few days of fuel available. Some may lack potable water and suffer shortages of medications, which are supplied on a daily basis to most facilities. Sanitation is a concern, and many employees are not reporting for work due to transportation or family issues.

## Nuclear Power Plants
Nuclear plants across the country experienced loss of off-site power and were required to shut down reactors. They are all equipped with emergency diesel generators to maintain critical functions and have over a week's fuel supply. However, with a widespread power outage, now they have very limited options for refueling. The federal government is facilitating fuel supplies to the nuclear power plants to avoid a Fukishima-type disaster.

## University Campuses
The administration and campus staff are working to find solutions. Although backup generators worked for critical systems, most of the water systems are offline, and dining halls have limited ability to prepare food. Campus officials meet to determine their course of action to care for the 20 million students attending colleges across the nation. Students are seen by some as a liability, by others as a resource. Many students head home if they have transportation available.

## Community Situations
Most citizens have enough nonperishable food and medication for about a week. Water heaters provide several days of potable water. Many are dependent on life-sustaining medications, which will be in short supply or unavailable. Stress from the disruption weighs on many with medical conditions as the accommodations that allow them to live a normal life at home are unavailable.

## Hospitals
As the power flickered, backup generators kicked in. Only emergency power and lighting is operational. Some pharmaceuticals are in short supply, and more patients are arriving as doctor offices and clinics are closed. Efforts to move severe patients to areas of the nation that are unaffected have put a huge strain on the entire nation's healthcare, as critical supplies are running low.

## Electric Utilities
Local electrical disruption managers are struggling with communications to respond to crew updates and status in the field. The nation has a limited number of large generators normally in reserve for peak loads, which are being pressed into service, but black starting some of these units is presenting huge challenges. Replacements for the damaged multi-megawatt generators have a manufacturing lead time of 18 months and depend on global supply chains. It will be months to years before the nation is back to full generation. This complicates the mission to keep fuel and food flowing and to sustain the U.S. economy. Partial restoration in areas of the country may be possible. Some islands of power exist, and a small percentage of renewable sources designed for "island operation" are generating limited localized supplies. Electric utility staff is working to restore power in areas as best as they can.

## Public Information

A small number of radio stations are on the air via backup generators, and are trying to locate community leaders to convey information to the public. They have managed to locate self-proclaimed experts who are predicting massive food and drug shortages, and recommending people take whatever measures they can to protect themselves. Public information officers struggle to craft messages for governors and mayors to relay to the public. Lack of power has complicated response efforts, and the massive scale of the outage has overwhelmed normal federal assets.

## DOD Issues

Base commanders have ordered lockdowns while they try to sort out the resources available to keep mission essential functions operating. Military installations usually rely on public utilities for the bulk of their power, with only critical buildings on base provided with backup power. A few with offline microgrid capability are better off with partial power, but are still coping with the lack of food, water, sewer, and other supplies. The food supply on base is provided by private contractors, with a one-week supply kept in stock, and a cache of Meals Ready to Eat (MREs) held in reserve. Water is provided to the base from the local water authority, base commanders' work through their respective chains of command to acquire resources outside the base from the larger community necessary to maintain readiness. Bases need to increase local security to discourage civilians from trying to access the base for resources that are not available.

## Agriculture

Farmers have a diverse group of problems. Some have milk cattle needing daily attention, but not enough power to last more than a few days. Although they may have plenty of food and water for family and the livestock, they worry that their rural area will be invaded by those looking for food. Many sleep in the paddocks guarding the livestock. The trucks that normally pick up the milk daily do not arrive. The nation is the breadbasket of the world, but the food is not where the population is, or in a form it can be used by most people. Without a ready supply of fuel and the logistics to manage the transportation, processing, packaging and distribution, the food that exists is not available to sustain the population.

Federal officials focus on distribution of process and packaged food, which is severely hampered by intermittent computer systems, Internet, and phone communications. Often shipments are sent, without knowing whether they arrived, or whether they were able to be distributed.

## Trucking

Trucking companies are working with their fuel suppliers to get a fuel delivery for their fleets. Many companies have contracts with food suppliers for distribution. Millions of people have evacuated their homes, especially from urban areas, attempting to find locations where electricity is available. Inoperable fuel pumps have resulted in many cars being abandoned along roads and highways, impeding essential transportation.

## Natural Gas and Fuels

Natural gas companies are trying to refurbish older natural gas-powered compressors to keep gas moving in the pipelines. Most pipeline compression had been converted to electric drive a few years earlier, but some are still functional. Reliable communication to govern pipeline operations is unavailable. Liquid fuel pipelines have stopped running, as pumps failed, or as refined supplies at the origination terminal stopped flowing. Tank farms have gone to a rationing system for remaining supplies, with public safety services and food production as primary recipients.

## Summary—Week Two

Most backup generators have run out of fuel supply in the affected areas. Some effort is being made to ship portable generators in from Texas and other unaffected areas, but the lack of a prioritization scheme hampers recovery efforts. Local governments are rationing the remaining fuel stocks to critical facilities, such as hospitals and government communications. Many communication systems are debilitated due to lack of generators and fuel, impeding emergency information distribution, and coordination of recovery efforts. Some jurisdictions are still supplying police, fire, and ambulance services, although many are beginning to address fuel, food, and supplies issues for the crews. Some microgrid communities or families are creating their own power. These families are trying to help neighbors, but fear they will be targeted for their food, fuel, generators, and other resources. The spirit of cooperation that prevailed during the first few days has diminished and continued panicked food purchasing is beginning to result in food shortages.

Medical supplies are being rationed, and the government is talking with the pharmaceutical companies to move production to open facilities and manage distribution. News of refinery operations, food packaging, and pharmaceutical manufacturing in areas with power helps to alleviate some concern about supply shortages to those with communication assets including shortwave radios and access to the remaining media.

Transportation is impaired due to large numbers of abandoned vehicles blocking roads. Train systems are functioning through verbal communications since the signaling systems are inoperable in much of the nation. The train systems are diverted to transporting bulk food and supplies to major population centers. Most federal agencies are still functioning. FEMA has already begun to move stored goods out of distribution warehouses to major population centers. As the magnitude of the event slowly unfolds, most Americans are almost totally reliant on the government for food and water, as their meager provisions are quickly exhausted. Cold settles in much of the nation, and those who live in rural areas accustomed to power outages are managing better than those in cities.

# End of Week Two

## Module One—Questions for Participants to Discuss in Small Groups

1. What issues would you and your family face?

2. What would your organization do after power has been out for over a week, two weeks?

3. At the end of week two, what things might you be doing to prepare for the continued lack of power?

Notes

# Cyberintrusion in the Electrical Sector
## Module Two
### Situation after 30 days

## Questions for Participants to Consider During This Module

1. If power was out for a month, how would you and your family cope?

2. What would life in your community look like?

3. What would your organization do without power for this length of time? Would people go to work? How would you pay them?

4. How would communications be maintained?

5. What kind of product and service exchange would take place between organizations?

6. How might you or your organization function under a Declaration of Emergency or martial law?

7. At the end of 30 days, what capabilities have you been able to restore?

8. How do you know what to expect in the future?


News of the magnitude of the event and the impact on the nation slowly moves through the nation through various means of communication (e.g., amateur radio users and occasional telephone service).

FEMA is working to get food supplies in the Midwest moved by rail to both coasts. Diesel fuel is the limiting factor. Power was restored in Cushing, Oklahoma, and has provided access to several million gallons of fuel stored there, that FEMA has commandeered. They are working with the FEMA regions to allocate scarce resources.

Elderly have started to die, due to lack of food, water, or life-sustaining medications. Millions with insulin or psychiatric illnesses are having difficulty coping with the hardships and are increasing effects on families, workforce, and environment.

Many responders have abandoned their positions to take care of their families. The Governor has requested support from the federal government to establish order within communities, support hospitals, and protect critical infrastructures.

Disposal of the bodies has become a major issue as the potential for disease necessitates action.

Nationwide, there is a massive delay in getting generators repaired or replaced. The handful of generator manufacturing companies is struggling and lack adequate materials from a global supply chain that depends on efficient transportation and communications. Transportation of new generation equipment onto ships and trains for delivery, which is normally a logistical challenge, has become even greater due to the fact that only a few U.S. ports are open and ships have had difficulty in accessing fuel due to the short supply. This challenge is magnified because much of new equipment weighs hundreds of tons and is physically large.

The extended blackout continues to impact numerous important functions at nuclear power plants (e.g., maintaining coolant flow, and storage and/or transportation of nuclear waste). The directors are working with the Nuclear Regulatory Commission, the Department of Energy, the Department of Defense Installation, Energy and Environmental Offices, the White House, and the Congressional Committees on Homeland Security, as well as with other nations.

Small hospitals have closed, combining resources with larger hospitals and the Department of Defense's medical community. All hospital operations are severely hampered by shortages of disposable personal protective equipment, disposable equipment, and medicines. Bodies placed in the morgues decay, and sanitation and disease outbreak have become serious concerns. Noncritical patients are sent back to their homes. The staff has set up makeshift triage anywhere they find it feasible. Providing supplies of food for the staff is now an additional challenge as the federal food stocks are depleted.

The grid operators continue to work to restore power to local generation plants using lower voltage transformers. Balancing the load is one of the prime issues and could cause greater damage. Restarting the power systems requires the grid operators to disconnect from the regional grid to get an island of power up and running. They are down to a skeleton crew, with only enough fuel for utility trucks to make short runs. Lack of fuel and staff are compromising the situation.

The radio stations have been unable to maintain power. They lowered the wattage of their transmitters and, by the end of the first week, most were off the air. Local officials had been trying to communicate via "reverse-911" phone dialers, although that system has been intermittent since the phone companies are inoperable in most places. Social order begins to break down as people run out of food and water. Fear of starvation drives many to loot in the larger cities and causes others to scavenge or barter. Civilian law enforcement has been enhanced or coopted by martial law.

Many citizens have taken to the road to look for resources in warmer climates or areas with power. Those with food and water have become more protective of their homes and families. Farmers slaughter some of their herds to use as barter for fuel or other food and, in some areas, to share with neighbors and community. Those in the north and northwest have taken to the woods to seek game to feed their families, but those resources are quickly depleted. A barter economy is developing throughout many of the communities with luxury supplies like toilet paper and weapons traded for food and water.

Four weeks into the event, there is some improvement in the electrical picture with some pockets of restored power, estimated at 35% restoration. The ability to obtain food, medicine, and clean water are hampered by a national logistical supply crisis. Water systems are only sporadically operational in the much of the nation. Food and pharmaceutical manufacturers have no access to raw materials or packaging supplies, and transport of final products are limited due to fuel shortage at either end—processing and delivery points.

## Summary—One Month

While generation plants continue to slowly come back online, repaired by cannibalizing or fabricating parts, most remain shut down due to generator replacements. Attempts to rewire the systems to lower transmission voltages are often nonproductive due to shortage of parts. The nation now has some power, but disruptions and brownouts are still the norm. In some areas of the northeast, there is sparse food distribution and shortages of fuel have made little progress in getting critical workforce to work, providing power to hospitals, running the government, operating schools, and performing other normal activities.

Communications systems are running for government emergency communications, although most of the nation is still without service. The count of lives lost is not known due to lack of communication to report. Sanitation has become a major issue and is being left to each individual community to contend. Burial of the dead is by hand in many areas and has become a priority for the community.

## One Month After the Event

### Module Two—Questions for Participants to Discuss in Small Groups

1. If power was out for a month, how would you and your family cope?

2. What would life in your community look like?

3. What would your organization do without power for this length of time? Would people go to work? How would you pay them?

4. How would communications be maintained?

5. What kind of product and service exchange would take place between organizations?

6. How might you or your organization function under a Declaration of Emergency or martial law?

7. At the end of 30 days, what capabilities have you been able to restore?

8. How do you know what to expect in the future?

Notes

# Notes

Notes

# IS YOUR CRITICAL NETWORK INFRASTRUCTURE AT RISK OF A SECURITY BREACH?

**YES!**

**WHY?**

**Hackers know how the network security game is played...**

## CHANGE THE GAME!

Avaya Fabric Connect offers solutions delivering network virtualization, cloud, mobility, and video—while simplifying the network, increasing deployment speed and building in higher levels of security.

Notes

# Cyberintrusion in the Electrical Sector
## Module Three

## Questions for Participants to Consider During This Module

1. Without power for three months, what would happen to you and your family? What actions would you take at this point?

2. What would life in your community look like? How would you work with your neighbors? Who would your neighbors be at this point?

3. How would communications be maintained?

4. What information would you need at this point?

Over the last month, conditions around the nation have sporadically improved. Citizens have been trying to cope with life as they now know it, but the magnitude of the situation is overwhelming. The ability to refine oil has been limited, and most sources of fuel have run out. Most communication devices remain out of commission in affected areas. People are often confined to communicating with those who are within walking distance of their current location.

Most nuclear power plants depleted their backup generator fuel supplies weeks ago. A few were able to reconnect to the grid and restart their reactors. They are serving as islands of power, and are large enough to black start other power generation. Other plants are challenged to keep the cooling systems online. Without coolant, the spent rods in storage will overheat, melt, and release radioactive contamination downwind. Plant procedures continue to be applied while the directors and staff members are in constant communication with the White House, Congress, NRC, DOE, DOD, and others. The local communities have taken it upon themselves to evacuate the area, and those remaining are being notified through town meetings of the status of the plant.

Small towns throughout the nation have been or are seeing a large fluctuation of close-by city dwellers moving into the area searching for food and water. There has been a migration of humanity to areas with power, causing a huge strain on every part of the nation. People in the medical community are starting to see a rise in various diseases due to inadequate sanitation, a food- and water-deprived population, and the continuation of cold weather.

## Summary—Three Months

Emergency generator on-site fuel supplies have been exhausted. Many areas are plagued by severe shortages of every kind. Citizens that had prepared for extended outages are running out of supplies. Millions are in a bad way. Millions have died and much of the nation feels as though it has been thrown back in time. The efforts of those fortunate enough to have electrical power, food, and water are thwarted by the lack of a reliable supply network. In many areas, violence erupted as roving gangs plundering remaining supplies from those that had prepared. There is no area of the nation left untouched. Fears of other nations taking advantage of this situation have not materialized, but remain in the national consciousness. As the nation slowly recovers power the damage to society remains.

# Three Months After the Event

## Module Three—Questions for Participants to Discuss in Small Groups

1. Without power for three months, what would happen to you and your family? What actions would you take at this point?

2. What would life in your community look like? How would you work with your neighbors? Who would your neighbors be at this point?

3. How would communications be maintained?

4. What information would you need at this point?

# Notes

# Notes

# Twelve Months After the Event

A year or longer outage could be caused by a large initial event or a series of multiple events over time. In addition, cascading effects could keep the power grid from full restoration.

## In the small groups discuss the following questions:

1.  What is the new normal for you, your community, and your organization?

2.  What happens based on experiences around the world when there is a long-term deprivation of resources?

Notes

# Questions After Participating in the Exercise

1. What changes can you make at home and in your organization to be more prepared?

2. What interactions do you want to have within your community including other organizations to be more prepared?

3. What if one community were able to make its own power:
   a. Would it become the emergency operations center for the state?
   b. How could it be protected?
   c. Should communities try to make their own power?

Notes

# Catastrophic Solar Storm Tabletop Exercise (TTX)

# Catastrophic Solar Storm

## Module One

### Questions for Participants to Consider During This Module

1. Since you could know that this is going to happen before the event, would that help you and your family be more prepared for this event? What would you do?

2. What would your organization do without power for two weeks?

3. At the end of week one, what things might you be doing to prepare for the continued lack of power?

## Scenario Background

*Adapted from "Managing Critical Disasters in the Transatlantic Domain—The Case of a Geomagnetic Storm" (February 23–24, 2010).*

Four days ago, an extremely large and complex sunspot cluster emerged on the sun. Several major solar flares erupted from this sunspot group two days ago.

Today, just as the large and complex region approaches center disk on the sun, a R5 (extreme) solar flare is detected with an associated with a coronal mass ejection (CME) by the NOAA Space Weather Prediction Center (SWPC) in Boulder, Colorado. This CME appears to be headed directly at the Earth and has a velocity consistent with the historically largest storms based on satellite observations. Coronal mass ejections drive geomagnetic storming at Earth and can cause significant electric power grid problems as a result. SWPC scientists begin modeling this coronal mass ejection to refine arrival time and initial forecast estimates of intensity state G4 (severe) to G5 (extreme) storming are possible, just 18–20 hours after initial detection of the eruption.

Electric power transmission, or "high-voltage electric transmission," is the bulk transfer of electrical energy, from generating power plants to substations. This is distinct from the local wiring between high-voltage substations and customers, called electric power distribution. Transmission lines, when interconnected with each other, become high-voltage transmission networks referred to in the United States as "power grids" or just "the grid." The United States has three major grids: the Western Interconnection, the Eastern Interconnection, and the Electric Reliability Council of Texas (or ERCOT) grid.

As further information comes in, SWPC confirms its initial analysis and states G5 (extreme) geomagnetic storming is expected just 18 hours after the initial eruption, with storming expected to persist for 36–48 hours after onset. NOAA SWPC issues a watch for G5 geomagnetic storming and initiates planning teleconferences with both NERC and FEMA in response. In addition to disrupting HF communications and GPS systems, G5 storming has the potential to disrupt bulk power grid, induce high voltages that can cause protective system problems, and damage EHV transformers (voltages higher than 345 kV) and sensitive control circuitry within various critical and noncritical systems. The NOAA spokesperson states this event has the potential to drive extreme geomagnetic storming, but that the intensity will remain uncertain until measurements of this CME are made by the Advanced Composition Explorer (ACE) spacecraft just upstream of Earth. At that time, higher confidence, short-term warnings will be issued. At this point, it is impossible to say whether this will be a once in five-year storm, once in 50, or once in 500. They estimate arrival time mid-morning on [today's date].

Public Information Officers across the nation are recommending people check their battery powered radios, flashlights, food supplies, and medicines. As a result, in several hours the metropolitan areas of the nation are out of essentials. In rural areas, people are largely relying on personal stocked inventory but are buying food, medicines, topping off fuel tanks, and loading up on feed for the animals.

The next update is issued just before 11 AM today as the CME strikes the ACE satellite, approximately 10–12 minutes before the storm is expected to impact the Earth's magnetosphere. The magnetic strength and orientation as measured by ACE confirm this is an extreme storm and G5 storming is largely a certainty. Utilities prepare to protect their systems and implement their plans.

## Impact Analysis—Day One

The storm creates massive ground-induced currents, overwhelming protective systems for transformers and electronic surge suppressors. Within minutes, the Western and Eastern Interconnection grids collapse. Some transformers in the Texas grid escape with minor damage. As the storm progresses, electrical and communication systems are disrupted around the globe.

FEMA, working from their National Response Coordination Center, attempts to collect damage assessment information, to assess the potential need for support to the nation. Widespread communication failures due to the loss of power to those systems hamper their efforts to get accurate information. Early estimates are that over 300 EHV transformers are heavily damaged, with many generation units damaged as well. While most generators are still viable, there is no way to transmit the power into the grid, with the EHV transformers out of service. Some outages were caused by recoverable protective systems tripping to protect the grid; however, many of the control systems that would normally reset these devices have been compromised.

Cell service is sporadic and Internet access is extremely limited given the overwhelming demand on these networks. Fiber optic cables have been spared damage, but many repeater stations are going offline as their backup power sources are exhausted.

Traffic signals are chaotic as normal sequencing halts and backup power sources are exhausted. Traffic in most metropolitan areas grinds to a halt.

The control and communication systems of metropolitan transit systems, like the one in New York City, have failed, leaving thousands of people trapped or stranded in the dark with no information on operation of subways and trains. Buses are still running, although many are stranded in gridlocked traffic jams. Transit agencies have minimal resources to address the situation and no communication with the trains or buses. Air travel is grounded.

High-frequency radio systems used by emergency management and amateur radio systems on "ham" radio frequencies are disrupted by the initial impacts of the solar storm as predicted by NOAA's SWPC to be unreliable during the first few days of the solar storm.

The Internet "E-Finance" is only partly operational and based on limited access to the Internet. Banks are closed due to the power outages. Businesses are not accepting checks nor debit or credit cards, as the transactions cannot be verified. The electronic-centric economy, as we know it, grinds to a halt.

# Summary—End of Day One

There are massive outages across much of the United States. Over 300 EHV transformers are believed to be out of service. Parts of the country were less affected, especially in the Texas grid system. A few transformers escaped damage areas of the Western and Eastern Interconnections, but more than 70% of the nation remains without power. Generators are expected to provide backup power to critical facilities for at least 72 hours, although most water utilities are down. Most cell service is spotty due to overuse and crowded circuits. Fuel supplies for backup generators typically will last for 72 hours.

Mass transit has shut down. Weather is relatively warm for December over much of the nation, thus heating is not a critical concern at this time. The extreme geomagnetic storm lasted for 24 hours ending on [two days after today's date]. The storm's intensity was similar to the Carrington Event of 1859, and full recovery of the U.S. power grid is expected to take years, due to shortages of replacement transformers and the cascading effects of the current outages. At this time, the Western and Eastern Interconnections are mostly inoperable, although the Electric Reliability Council of Texas (ERCOT) grid is still functioning. The status of Canadian's grid is partially functional, although unable to assist with the Eastern Interconnection at this time. The Mexican grids are completely inoperable, and they are reaching out to ERCOT for assistance.

# End of Week Two (depending on the specific location, this may be 10–14 days)

## Communications

High-frequency radio communications are back to normal in most of the United States after the solar storm is over. In most of the metropolitan areas and larger towns, however, many of the transformers have suffered damage, making power unavailable. Backup transmitters are deployed and operating on generator power as long as backup fuel is available. Most phone systems have avoided direct damage, but are beginning to fail due to lack of replenishing of fuel farms supplying diesel generators. Broadband communications begin to fail as backup power sources are spotty.

## Care Facilities

Care facilities providing care to several million elderly and disabled have lost power to critical equipment and systems. Backup generators kick in, and most facilities have a few days of fuel available. Some lack potable water as well as shortages of medications, which are supplied on a daily basis to most facilities. Sanitation is a concern, and many employees are not reporting for work.

## Nuclear Power Plants

Nuclear plants across the country experienced loss of off-site power and are required to shut down their reactors. They are all equipped with emergency diesel generators to maintain critical functions and have limited fuel supply and, with a widespread power outage, now they have very limited options for refueling. Many plants suffered GIC damage to generator step-up (GSU) transformers weighing hundreds of tons, control systems, and many electrical or electronic components. Staff are working to conduct damage assessments and to develop a strategy to recover their systems.

## University Campuses

The administration and campus staff members are working to maintain life safety. While the backup generators are kicking in for critical systems, most of the water systems are offline, and dining halls have limited ability to prepare food. Campus officials meet to determine their course of action to care for the 20 million students attending colleges across the nation. Students are seen by some as a liability, by others as a resource. Many students head home if they have transportation available.

## Community Situations

Most citizens have enough nonperishable food and medication for about a week. Water heaters provide several days of potable water. Many are dependent on life-sustaining medications, which will be in short supply or unavailable. Stress from the disruption weighs on many with medical conditions as the accommodations that allow them to live a normal life at home are unavailable.

## Hospitals

Hospitals had sufficient warning to be prepared for a power outage. Most critical surgeries had been completed, and elective surgeries were postponed. As the power has flickered, backup generators have kicked in. Elevators are limited and only emergency power and lighting is operational. Some pharmaceuticals are in short supply, and more patients are arriving as doctors' offices and clinics are closed.

## Electric Utilities

Local electrical disruption managers restored more reliable radio communications to respond to crew updates and status in the field. They are working to re-establish broadband communications to determine the extent of damage to the system, particularly the high-voltage transformers that link the grid to the distribution systems. Reports of smaller transformers exploding have reached the office, but have not been verified.

The nation has a very limited supply of replacement EHV transformers. The transformers serving the bulk power grid (above 230 kV) have a manufacturing lead time of 18 months and, currently, are usually built outside the United States. If one of these countries has been affected by the solar flare event, then the lead time for this transformer may be longer than 18 months. A few modular units owned by the federal government can be trucked on-site and assembled as a temporary solution.

With much of the world's power supply disrupted, the need for replacement transformers has created backup of demand, which would take 30 years to produce at current world production capacity. Without sufficient electricity, converting other facilities to produce the transformers would be painstaking. Most EHV transformers are built outside the United States, and the likelihood of this country being first in line for units is low. Further complications involve the shortage of skilled technicians to build transformers or to repair existing systems.

Based on the shortage of replacement EHV transformers, full restoration of the grid will be measured in years, not months. This complicates the mission to keep fuel and food flowing and to sustain the economy. Partial restoration in areas of the country may be possible. Some islands of power exist, and some renewable sources not connected to the grid are generating limited localized supplies.

## Public Information

Some radio stations are on the air via backup generators, and are trying to locate community leaders to convey information to the public. They have managed to locate self-proclaimed experts who are predicting massive food and drug shortages, and recommending people take whatever measures they can to protect themselves. Public information officers struggle to craft messages for governors and mayors to relay to the public. Lack of power had complicated response efforts, and the massive scale of the outage has overwhelmed federal assets.

## DOD Issues

Base commanders have ordered lockdowns while they try to sort out the resources available to keep mission-essential functions operating. Military installations usually rely on public utilities for the bulk of their power, with only critical buildings on base provided with backup power. A few with offline microgrid capability are better off with partial power, but are still coping with the lack of food, water, sewer, and other supplies. The food supply on base is provided by private contractors, with a one-week supply kept in stock, and a cache of MREs held in reserve. Water is provided to the base from the local water authority. Through their respective chains of command, base commanders work to acquire necessary resources from the larger community outside the base to maintain readiness. Bases need to increase local security to discourage civilians from trying to access the base for resources that are not available.

## Agriculture

Farmers have a diverse group of problems. Some have milk cattle needing daily attention, but not enough power to last more than a few days. Although they may have plenty of food and water for family and the livestock, they worry that those looking for food will invade their rural areas. Many sleep in the paddocks guarding the livestock. The trucks that normally pick up the milk daily do not arrive. Farmers are trying to complete harvesting in southern states with their remaining fuel. Storage is an issue, with many just dumping it in huge piles. Although the nation is the breadbasket of the world, the food is not where the population is, nor is it in a form that can be used by most people. Without a ready supply of fuel and the logistics to manage the transportation, processing, packaging, and distribution, the food that exists is not able to sustain the population.

Federal officials focus on distribution of process and packaged food, which is severely hampered by lack of computers, Internet, and phone communications. Often, shipments are sent without knowing whether they arrived, or whether they could be distributed.

## Trucking

Trucking companies are working with their fuel suppliers to get a fuel delivery for their fleets. Many companies have contracts with food suppliers for distribution. Millions of people have evacuated their homes, especially from urban areas, attempting to find locations where electricity is available. Inoperable fuel pumps have resulted in many cars being abandoned along roads and highways, impeding essential transportation.

## Natural Gas and Fuels

Natural gas companies are trying to get older natural gas-powered compressors to keep gas moving in the pipelines. Many had been converted to electrical power a few years ago, but some are still functional. They have no reliable communication to know whether the gas will remain flowing into the pipeline. Other liquid fuel pipelines have stopped running, as pumps failed, or refined supplies at the origination terminal stopped flowing. Tank farms have gone to a rationing system for remaining supplies, with emergency services and food production as primary recipients.

## Summary—End of Week Two

Approximately 300 HV transformers are out of service nationwide, with some utilities able to repair some units with minor damage. Most backup generators have run out of fuel supply. Local governments are rationing the remaining stocks to critical facilities, such as hospitals and government communications. Some jurisdictions are also able to supply police fire and ambulance services, although many are beginning to face fuel, food, and supplies issues for their crews. Some microgrid communities or families are creating their own power. These families are trying to help neighbors, but fear they will be targeted for their power supplies.

Power is only reliably available in the Texas grid. Efforts to keep refineries, food distributors, and pharmaceutical manufacturers producing help alleviate some concern about food, fuel, and water shortages. Medical supplies are being rationed, and the government is talking with the pharmaceutical companies to get warehoused supplies. Phone services are generally inoperable outside of the Texas grid areas.

Train systems are largely functioning through verbal communications since the signaling systems are inoperable within the Western and Eastern Interconnections. The train systems are diverted to transporting bulk food and supplies to major population centers. Most federal agencies are still functioning to provide critical services.

FEMA has begun to move stored goods out of distribution warehouses to major population centers. As the magnitude of the event slowly unfolds, most Americans are almost totally reliant on the government for food and water, as their limited provisions are quickly exhausted.

## End of Week Two

### Module One—Questions for Participants to Discuss in Small Groups

1. Since you knew that this was going to happen before the event, did that help you and your family be more prepared? What did you do to prepare?

2. What is your organization doing without power for these two weeks?

3. At the end of week one, what things might you be doing to prepare for the continued lack of power?

# Notes

# Catastrophic Solar Storm
## Module Two

### Questions for Participants to Consider During This Module

1. If power was out for a month, how would you and your family cope?

2. What would life in your community look like?

3. What would your organization do without power for this length of time?  Would people go to work?  How would you pay them?

4. How would communications be maintained?

5. What kind of product and service exchange would take place between organizations?

6. How might you or your organization function under a Declaration of Emergency or martial law?

7. At the end of 30 days, what capabilities have you been able to restore?

8. How do you know what to expect in the future?


News of the magnitude of the event has slowly spread through the nation through various means of communication (e.g., amateur radio users and occasional telephone service). Those with power are nearly as affected as those without, by the influx of hundreds of thousands of people seeking food, shelter, and medicine, overwhelming the resources that exist.

FEMA is working to get food supplies in the Midwest moved by rail to both coasts. Diesel fuel is the limiting factor. Power was restored in Cushing, Oklahoma, and has provided access to several million gallons of fuel stored there, which FEMA has commandeered. They are working with the FEMA regions to allocate scarce resources.

Elderly people are dying due to lack of food, water, or life-sustaining medications. Millions with insulin dependence or psychiatric illnesses are having difficulty coping with the hardships and are increasing effects on families, workforce, and environment.

Many responders have abandoned their positions to take care of their families. The governors have requested support from the federal government to establish order within communities, support hospitals, and restore critical infrastructures.

Disposal of the bodies has become a major issue as the potential for disease necessitates action. Freezing weather over much of the nation complicates burial, but prevents much of the potential for disease.

Worldwide, there is a massive delay in getting transformers repaired or replaced since normal lead times are over a year. Backlogs can make that several years even if additional factory capacity is begun. A handful of transformer manufacturing companies are struggling for production volume because some of their operating systems are nonfunctional due to electrical failure, and others lack adequate raw material sources. All the major countries are vying for position in the very small production volume. Transportation of the units onto ships and trains for delivery, which is normally a logistical challenge, has become even more so due to that fact that only a few southern U.S. ports are open, and ships have had difficulty in accessing fuel due to the short supply.

The extended blackout across the Western and Eastern Interconnections continues to impact numerous important functions at nuclear power plants (e.g., maintaining coolant flow, and storage and/or transportation of nuclear waste). The directors are working with the Nuclear Regulatory Commission, the Department of Energy, the Department of Defense Installation, Energy and Environmental Offices, the White House, the Congressional Committees on Homeland Security, and with other sovereign nations.

Small hospitals have closed combining resources with larger hospitals and the Department of Defense's medical community. All hospital operations are severely hampered by shortages of disposable personal protective equipment, disposable equipment, and medicines. Bodies placed in the morgues decay, and sanitation and disease outbreak have become serious concerns. Noncritical patients were sent back to their homes. The staff has set up makeshift triage. Providing supplies of food for the staff is now an additional challenge.

The grid operators continue to work to restore power to local generation plants using lower voltage transformers. Balancing the load is one of the prime issues and could cause greater damage. Restarting the power systems requires the grid operators to disconnect from the regional grid to get an island of power up and running. They are down to a skeleton crew, and with only enough fuel for utility trucks to make short runs. Lack of reliable communications, fuel, staff, and food are compromising the situation.

The radio stations have been unable to maintain power. They lowered the wattage of their transmitters and, by the end of the first week, most were off the air. Local officials had been trying to communicate via "reverse-911" phone dialers, although that system has been intermittent since the phone companies are inoperable in most places. Social order begins to break down as people run out of food and water. Fear of starvation drives many to loot and causes others to scavenge or barter. Civilian law enforcement has been enhanced by martial law.

Many citizens have taken to the road to look for resources in warmer climates. Those with food and water have become more protective of their homes and families. Farmers slaughter their herds to use as barter for fuel or other food and, in some areas, to share with neighbors and community. Those in the North and Northwest have taken to the woods to seek game to feed families. A barter economy is flourishing throughout many of the communities with luxury supplies like toilet paper and weapons traded for food and water.

Four weeks into the event, there is some improvement in the electrical picture with some pockets of restored power in the Eastern Interconnection, estimated at 25% restoration. The Western community power availability is still at 10%, while the Texas Interconnection is 80% operational. Restoration has scored some successes but, for every success, many challenges remain.

The ability to obtain food, medicine, and clean water are hampered by a national logistical supply crisis. The "just-in-time" system, that works so well to provide low-cost goods, has become the bane of the nation's ability to recover. Food and pharmaceutical manufacturers have limited access to raw materials or packaging supplies, and transport of final products are limited due to fuel shortage at either end—processing and delivery points. Many water systems in the Texas grid are operational, but are only sporadically operational in the other regions.

## Summary—One Month

As power systems continue to slowly come back online, repaired by cannibalizing or fabricating parts, most remain shut down due to transformer replacements. Attempts to rewire the systems to lower transmission voltages are nonproductive due to shortages of parts. Estimates are that the West Interconnection has 10% of power online, the Eastern Intersection has 25% power online, and TX has 80% online. So the nation now has some power, but disruptions and brownouts are still the norm. There is sparse food distribution, and shortages of fuel have made little progress in getting critical workforce to work, providing power to hospitals, running the government, operating schools, and performing other normal activities.

Communication systems are running for government emergency communications, although most of the nation is still without service. The count of lives lost is not known due to lack of communication among the workforce to report. Sanitation has become a major issue and is being left to each individual community to contend. Burial of the dead is by hand in many areas and a priority for the community.

# One Month After the Event

## Module Two—Questions for Participants to Discuss in Small Groups

1. If power was out for a month, how would you and your family cope?

2. What would life in your community look like?

3. What would your organization do without power for this length of time? Would people go to work? How would you pay them?

4. How would communications be maintained?

5. What kind of product and service exchange would take place between organizations?

6. How might you or your organization function under a Declaration of Emergency or martial law?

7. At the end of 30 days, what capabilities have you been able to restore?

8. How do you know what to expect in the future?

Notes

# Notes

# Notes

*A Cyber Innovation Labs Venture*

# Max-Survivability Center

## Differentiation

- First commercially available Protected Platform from EMP/HEMP/IEMI and GMS events

- 360° tested, certified and shielded Tier 3+ Data Center rating

- Long-life survivability for mission critical apps & services

- SSAE16 SOC 2, PCI and HIPAA/HITECH compliance

- Value engineered at price points of traditional, legacy market offerings

- Real-dollar SLA offers $50K - $10M Uptime Guarantee

Enterprise-class data centers require sophistication in design, management, compliance, scalability and **Electromagnetic Pulse (EMP)** and **Geomagnetic Storm** protection to support a wide variety of client IT needs. The EMP Resilient Facility solution, coined by EMP GRID Services, leverages best-of-breed technologies to furnish enterprises with complete access, managed data capabilities and EMP hardening across the entire communications infrastructure.

### EMP Resilient Facility Delivers

- Tier 3+ availability, uncompromising security, reliability, and control supported by industry-leading SLAs.
- **Highest level of compliancy - PCI, HIPAA, HITRUST, SSAE 16 SOC 1 and SOC 2.**
- Six-tiered security perimeters, fully monitored 24x7, with live on-site support.
- Customizable and scalable solutions to ensure ongoing business support and reliability.
- Custom designed EMP/HEMP enclosures and shelters.
- 100% Private, single-tenant Cloud solution offers unparalleled performance and uptime (99.99%).

### EMP Resilient Facility Functions

- 360° tested, certified and shielded Tier 3+ Data Center platform.
- All MEP/FP (electrical, mechanical and fire protection) stand-alone with 2N redundancy.
- **$50K - $10M SLA guaranteed uptime performance per customer.**
- Primary emergency generation (diesel / natural gas / fuel cell) within shielded containment supported by **30 days of uninterrupted on-site fuel storage**.
- Complete telecom infrastructure contained and protected with mobile rollout deployment.
- Mirrored and fully managed EMP/Geomagnetic Storm protected recovery facilities.

EMP GRID Services | www.empgridservices.com | P. 312.593.6106

Notes

# Catastrophic Solar Storm
## Module Three

## Questions for Participants to Consider During This Module

1. Without power for three months, what would happen to you and your family? What actions would you take at this point?

2. What would life in your community look like? How would you work with your neighbors? Who would your neighbors be at this point?

3. How would communications be maintained?

4. What information would you need at this point?

Over the last month, conditions around the nation have grown progressively worse without an end in sight. Emergency generator on-site fuel supplies have been exhausted in most places. Citizens have been trying to cope with life as they now know it, and the magnitude of the situation is overwhelming. The ability to refine oil has been decimated, and most sources of fuel have run out. Cellphones and many other communication devices are out of commission across the nation. People are often confined to communicating with those who are within walking distance of their current locations.

Most nuclear power plants have depleted their backup generator fuel supplies weeks ago. Other plants are challenged to keep the cooling systems online. Without coolant, the spent rods in storage will overheat and release radioactive contamination, or explode. Plant procedures continue to be applied while the director and staff members are in constant communication with the White House, Congress, NRC, DOE, and others. The local communities have taken it upon themselves to move out of the area, and those remaining are being notified through town meetings of the status of the plant.

Like Apollo 13, the nuclear operators and engineers work to come up with innovative solutions as best they can. Small towns throughout the nation are seeing a large increase of close-by city dwellers attempting to move into rural areas in search of food, water, and other resources. In many cases, this influx leads to conflict over dwindling basic resources. Multiple diseases are breaking out due to inadequate sanitation, food and water deprivation, and the effects of cold weather.

## Summary—Three Months

The nation experienced a bitter winter. Many areas are plagued by severe shortages of every kind. Citizens that had prepared for extended outages are running out of supplies. Millions are in a bad way. Millions have died and much of the nation feels as though it has been thrown back in time. The efforts of those fortunate enough to have electrical power, food, and water are thwarted by the lack of a reliable supply network. Survivor guilt plagues those fortunate enough to have made it through the winter.

# Three Months After the Event

## Module Three—Questions for Participants to Discuss in Small Groups

1. Without power for three months, what would happen to you and your family? What actions would you take at this point?

2. What would life in your community look like? How would you work with your neighbors? Who would your neighbors be at this point?

3. How would communications be maintained?

4. What information would you need at this point?

# Notes

# Notes

# Twelve Months After the Event

A year or longer outage could be caused by a large initial event or a series of multiple events over time. In addition, cascading effects could keep the power grid from full restoration.

## In the small groups discuss the following questions:

1. What is the new normal for you, your community, and your organization?

2. What happens based on experiences around the world when there is a long-term deprivation of resources?

Notes

# Questions After Participating in the Exercise

1.  What changes can you make at home and in your organization to be more prepared?

2.  What interactions do you want to have within your community including other organizations to be more prepared?

3.  What if one community were able to make its own power:

    a. Would it become the emergency operations center for the state?

    b. How could it be protected?

    c. Should communities try to make their own power?

# Notes

# Notes

# High-Altitude Electromagnetic Pulse Attack
# Tabletop Exercise (TTX)

# High-Altitude Electromagnetic Pulse
## Module One

### Questions for Participants to Consider During This Module

1. If there were a no-notice event, what would you and your family do?

2. How would you get communications to know what happened?

3. What would your organization do to manage this incident?

4. At the end of week one, what things might you be doing to prepare for the continued lack of power?

## Scenario Background

At 10:35 am, the power went out in much of the nation. Hundreds of in-flight civilian aircraft failed and crashed. A few radio and television stations are broadcasting outside the affected area, which covers most of the eastern and middle of the nation. Only the far southwest seems to be spared. Cellular service is out across much of the nation, although many phones are still operational. Cell providers are attempting to provide priority service and determining whether they can restore service. Landline telephone and Internet services are unavailable due to damaged equipment at network operation centers and central offices. The electric power grid is inoperable due to damaged equipment at control centers and substations across the country including generators, transformers, and SCADA systems. Aviation and military authorities report that a high-altitude nuclear blast has occurred. In-person reports are the key mechanism for communication. Some vehicles appear to be out of service, although many are still operating, especially those parked in underground garages. The stalled vehicles and the lack of traffic control infrastructure gridlock cities and major thoroughfares.

### Impact Analysis—Day One

### Power
Homes and businesses across most of the nation are without power.

### Water
Loss of water pressure in most metropolitan areas is felt immediately. A few water systems have operational backup generators and may continue to operate for a few days, until fuel supplies run out. Water pumping station electronic SCADA systems are also damaged, introducing further delays in system restoration.

### Sanitation
Sewage systems without power back up and flow out of toilets, septic tanks, and sewers in low-lying areas. In search of water, people turn to sources of groundwater, spreading typhoid, and other diseases spread from sewage-induced groundwater contamination.

### Food
Refrigerated and frozen foods need to be consumed within one day. Nonperishable food is safe to eat but, because of a run on food supplies and lack of restocking, most homes and supermarkets, restaurants, food pantries, etc. are out of food within a day.

## Financial Services

The nation's financial systems shut down. Retail stores close. Cash and barter are the only options to trade, with barter replacing cash as goods owners realize the severity of the outage. Some areas of the nation have power and some hardened data centers continue to operate. However, the bulk of the nation's Internet service that is essential for financial transactions is inoperable due to lack of power and EMP damage to network operation centers.

## Fuel

The power outage means that most gasoline and diesel fuel cannot be pumped out of storage. Most natural gas compressors for pipelines that have recently been converted from natural gas to electric-driven compression to comply with air-quality standards fail. Latent pressure in gas pipelines keeps gas flowing for a day or two. Many backup generators are inoperable due to the E1 pulse.

## Communications

Most landline phones fail immediately. One or more of the cellphone companies with computer control systems and backup generator controls that happened to survive the initial attack continue to work for a few hours. As tower battery backups and diesel fuel run out; however, these systems also fail. Failure of long-haul systems prevents long-distance telephone service and Internet connectivity. Most satellite phones continue to function. Some radio and television stations with EMP-protected backup power may continue to operate. Many ham stations continue to function, especially the older, tube-operated units or those that were not plugged into antennae and power systems at the time of the event. Some emergency radio dispatch systems continue to function where equipped with protected backup power. There is no Internet service due to failures of major network operation centers and data centers. Electric power control centers are similarly affected, leaving operators without situational awareness, and thus unable to assist with grid restoration by monitoring and/or controlling the state of the grid.

## Transportation

Although many planes, trains, and automobiles are operable, a combination of a lack of fuel (and stalled cars) and all transportation routing and monitoring systems (NAVAIDS) being down (no air traffic control or radar for planes, track switching for trains, or stoplights for cars) means that transportation is gridlocked and only minimally functional for some days, then failing completely as fuel supplies are exhausted. A focus on clearing major routes keeps fuel trucks and critical supplies moving to support population centers for a short time. Resourceful people drain fuel from inoperable vehicles to supply their needs.

## Government/Emergency Services

Protected backup power allows some governmental services to function for a short time, but command or control communications are very limited after a few days. Satellite phones continue to function. Hardened FEMA and military communication systems continue to function. Personnel availability becomes an issue, as people stay with their families to deal with the crisis, or have no transportation to get to work.

## Healthcare

Only hospitals with functioning backup power operate for a couple of days. After that, most vulnerable patients die, many medicines expire due to lack of refrigeration, and healthcare is basic at best. With no water and power for electricity-intensive hospital sanitation, any hospital struggling to continue to operate becomes unusable. Most have plans to evacuate in the event of a long-term power outage.

## Workforce/Commerce

Some businesses try to stay open for a few days, using cash. After that, businesses in the affected areas close. Some business owners try to protect their stock from looting. Barter predominates for commerce.

## Security

With communication limited, and transportation largely gridlocked, police and National Guard effectiveness are severely hampered. After days, lack of food and water make personnel retention extremely difficult. The concern for their families' safety drives many personnel to abandon what they see as a hopeless situation.

## Nuclear Power

Nuclear power plants in the blackout area trip offline and are shut down. Nuclear facilities have around a week of backup diesel fuel, so the core and the spent-rod cooling ponds can be kept from melting or burning for around a week, where the backup generators are operational. Many close to large supplies of water do what they can to divert water for cooling purposes. Electronic controls on many diesel generators are damaged such that portable replacement generators are needed. Portable diesel generators are organized and convoyed into the affected area for disabled plants. In some cases, control system failures also make maintaining cooling systems difficult. Back-up fire hose cooling does not work, since the city water system is down, where adequate fuel supplies cannot be trucked in from remote locations.

## Summary—Day One

FEMA, working from their National Response Coordination Center, attempts to collect damage assessment information, to assess the potential need for support to the nation. Widespread communication failures hamper their efforts to retrieve accurate information. Early estimates are that over 300 EHV transformers are severely damaged, with many generation units damaged as well. Although most generators are functional themselves, repair of their electronic control systems introduces several weeks of delay in restarting them. However, it will take longer to transmit the power into the grid, with the grid control centers, SCADA systems, and EHV transformers disabled. Some outages were caused by recoverable protective systems tripping to protect the grid. However, many of the control systems that would normally reset these devices have been compromised.

## Week Two

There are massive outages across the entire Eastern Interconnect. The Western and Texas grid systems remain operable. The status of Canada's grid is partially functional, although unable to assist with the Eastern Interconnection at this time. Generators provide backup power to many critical facilities for only 72 hours. Consequently, most water utilities are down. Most phone services are disrupted including cell service due to the EMP effect on copper lines and long-haul optical fiber repeaters. Most backup generators are not operational. With extremely limited power restoration across the nation, by the end of the first week, many critical infrastructures remain inoperable, and some nuclear power plants become radioactive contamination disaster areas.

### Communications

Commercial telephone and Internet service remains disabled. Ham radios and military EHF satellite communication and a limited number of local police, fire, and private LMR systems are still in service. One local cellular provider has very limited capacity. One radio station continues to broadcast. No television stations are on the air in the affected area.

### Nuclear Power Plants

Backup fuel and generators for nuclear power plants have been airlifted in and, although all nuclear power plants within the affected area are shut down, their cores and cooling ponds are operable.

### Power

Bulk power system remains inoperable to most of the nation. Operational power systems are struggling to en-

ergize some generators near the boundaries of the interconnects, but within the most of nation, the electrical system has not been able to black start. Electric power control center electronics in exposed regions have suffered burnout, leaving operators unable to monitor and control the state of the grid or coordinate the black-start process. Many generators have not been restarted due to SCADA system damage and grid connectivity problems.

## Fuel

Fuel deliveries are nonfunctional within several days. Refining and transportation capabilities remain out of service.

## Water and Sanitation

Much of the nation is without water and basic sanitation. Public health is in jeopardy from diseases like typhoid and cholera. The cold is slowing the spread of diseases.

## Food

Most families have about a week's supply of food. Federal stockpiles are not sufficient to supply the nation, and distribution channels are unavailable. Focus is on supplying the military, emergency services, and government officials to try to maintain continuity of government.

## Financial Services

Financial services are nonfunctional due to Internet and local area network damage. Barter becomes the primary mode of trade, with point of gun as an alternative.

## Transportation

Vehicle transportation within the effected region is limited to emergency vehicles (with some individuals that have fuel in their cars) and government approved flights and rail to carry in relief supplies. Primary transportation is by foot or bicycle, with mass migration away from cities clogging roadways and creating demands for sheltering needs for large populations.

## Government/Emergency Services

Enormous public concerns and some outbreaks of looting gangs.

## Healthcare

With no power for electricity-intensive hospital sanitation, many hospitals struggle to continue to operate, or have become unusable. The military works to establish field hospitals, but are hampered by lack of fuel, electricity, and water/sanitation as well. Without healthcare, food, water, and sanitation, many face death from starvation and other deprivation in the cold weather.

## Workforce/Commerce

The economy now is limited to some cash transactions and barter. Most businesses have closed.

## Security

Police and even National Guard effectiveness are severely hampered. Personnel retention remains extremely difficult. Looting has begun, and threatens to become the source of social breakdown as food, water, and other resources have run out.

# Summary—Week Two

A massive external effort is being made to deal with the crisis. The basic problem is that recovery capabilities are not designed to deal with such massive numbers of people and large areas where all utilities are simultaneously down.

It only has been possible to get power to some areas of the nation bordering existing operable systems. The EMP damaged most islanded power and alternative systems. Many backup generators are inoperable due to EMP effects on their electronic control systems. Operational emergency generators and fuel have been brought in by helicopter to some critical areas. External military capability is being used to secure and provide critical cooling fuel to nuclear power plants.

In many urban areas, it soon became obvious the best course of action was evacuation, requiring large refugee camps to be set up in areas with power. Main roads were cleared, and continuous convoys of food and water trucks were tried in some areas, but quickly became overwhelmed by the sheer numbers. A few relief ships have been sent with supplies to port cities on the Eastern Seaboard, although there are not nearly enough to supply needed food, water, and emergency supplies.

The military is concerned about foreign intervention, and must weigh what resources can be diverted to support civilian needs. Civilians come to their gates hoping for access to food, water, and electric power from base generators, but are turned away. If the government can determine who is responsible for the EMP attack, they would plan a retaliatory strike.

Emergency facilities are being set up in all the key refugee sites. Transportation is being provided to ferry people by rail and road (outside the affected area) to other cities, which also are setting up emergency shelters. Areas with power have been inundated with refugees and are struggling to provide for their needs. Other countries are being asked to bring in assistance, which may take weeks to arrive, due to logistical difficulties. Many Americans look to emigrate to other nations that are not affected by the EMP incident.

# END OF WEEK TWO

## Module One—Questions for Participants to Discuss in Small Groups

1. A no-notice event occurred. What are you and your family doing at this point?

2. How are you getting communications to know what happened?

3. What is your organization doing to manage this incident?

4. At the end of week one, what things might you be doing to prepare for the continued lack of power?

Notes

# High-Altitude Electromagnetic Pulse
## Module Two

### Questions for Participants to Consider During This Module

1. If power was out for a month, how would you and your family cope?

2. What would life in your community look like?

3. What would your organization do without power for this length of time? Would people go to work? How would you pay them?

4. How would communications be maintained?

5. What kind of product and service exchange would take place between organizations?

6. How might you or your organization function under a Declaration of Emergency or martial law?

7. At the end of 30 days, what capabilities have you been able to restore?

8. How do you know what to expect in the future?

News of the magnitude of the event has slowly spread across the nation through various means of communication and reporting within the affected area (e.g., amateur radio users and occasional telephone service, mostly word of mouth, with associated rumors and false information).

FEMA is working to get food supplies in the Midwest moved by rail to the East Coast. Diesel fuel is the limiting factor. FEMA is working with the FEMA regions to allocate scarce resources to be directed to the affected areas. The magnitude of the crisis overwhelms the resources available and the logistical and communications challenges make the efforts seem futile.

Many elderly have died, due to lack of food, water, or life-sustaining medications. Millions with insulin or psychiatric illnesses are having difficulty coping with the hardships and are magnifying the effect of the incident on families, workforce, and environment. Some are committing suicide, rather than face starvation or disease. Burial of the dead has become an important task to provide closure to those still living, and to prevent further spread of disease. Millions begin to face starvation and the intense cold.

Many responders have abandoned their positions to take care of their families. Governors have requested support from the federal government to establish order within communities, support hospitals, and protect critical infrastructures. Unfortunately, most federal workers are facing the same challenges, transportation problems, and family concerns as the rest of the nation.

Nationwide, there is a massive delay in getting transformers and generators repaired or replaced. The handful of transformer and generator manufacturing companies are struggling with production capability because some of their operating systems are nonfunctional due to electrical failure, and others lack adequate raw material sources. All the major transformer/generator-producing countries are attempting to increase their production volume, but this is difficult because of the hand-made and custom-built nature of transformers. Transportation of the units onto ships and trains for delivery, which is normally a logistical challenge, has become even more so due to that fact that only a few southern U.S. ports are open, and ships have had difficulty in accessing fuel due to the short supply.

The extended blackout across much of the nation continues to impact numerous important functions at nuclear power plants (e.g., maintaining coolant flow, and storage and/or transportation of nuclear waste). The directors are working with the Nuclear Regulatory Commission, the Department of Energy, the Department of Defense Installation, Energy and Environmental Offices, the White House, the Congressional Committees on Homeland Security, and with other sovereign nations.

Most hospitals have closed due to lack of water, waste treatment, electric power, medicines, personal protective equipment, and other medical supplies. Bodies placed in the morgues decay, and sanitation is a serious concern. Formerly minor injuries become life-threatening.

Grid operators have restored most grid monitoring control centers. Many damaged SCADA systems have been repaired or replaced. The electric utilities are working to restore power to local generation plants using lower voltage transformers. Balancing the load is one of the prime issues and could cause greater damage. Restarting the power systems requires the grid operators to disconnect from the regional grid to get an island of power up and running. Grid operators are down to skeleton crews, and with only enough fuel for utility trucks to make short runs. Lack of fuel, food, adequate communications, and staff are complicating the situation.

The radio stations have been unable to maintain power. They lowered the wattage of their transmitters, but by the end of the first week, most were off the air. Local officials had been trying to communicate via "reverse-911" phone dialers, though that system has been intermittent since the phone companies are inoperable in most places. Social order has broken down, as people have run out of food, water, and heat. Fear of starvation drives many to loot and causes others to scavenge or barter. Law enforcement has been replaced by martial law.

Many citizens, who are able, are taken to the road to look for resources. Those with food and water have become more protective of their homes and families. Farmers slaughter their herds to use as barter for fuel or other food and in some areas to share with neighbors and the community. Many people in more rural areas have taken to the woods to seek game for feeding families. A barter economy is developing throughout many of the communities with what were traditionally luxury goods like gold jewelry being traded for supplies like toilet paper and ammunition traded for food and water.

Four weeks into the event, there is some improvement in the electrical picture, with some pockets of restored power in the nation estimated at 25% restoration. This has not been sufficient to provide significant relief, as the logistical supply chain is still broken, financial markets are not working, and food and medicine are not available on any consistent basis.

The ability to obtain food, medicine, and clean water are hampered by a national logistical supply crisis. Food and pharmaceutical manufacturers have no access to raw materials or packaging supplies, and transport of final products are limited due to fuel shortage at either end. Maintaining fuel supplies to nuclear power plants is a national priority, especially for the cooling of spent fuel rods. Fukishima-like radiation contamination events could occur.

## Summary—One Month

Although power systems continue to slowly come back online, repaired by cannibalizing or fabricating parts, most remain shut down due to transformer and command and control failures. Attempts to rewire the systems to lower transmission voltages are often nonproductive due to shortages of parts. Estimates are that the nation has 25% power online. There is sparse food distribution. Shortages of fuel have made little progress in getting critical workforce to work, providing power to hospitals, running the government, operating schools, and so on.

Communications systems, if available, are being used for government emergency communications, although most of the nation is without service. The count of lives lost is not known due to lack of communication among the workforce to report. Sanitation has become a major issue and is being left to each individual community to contend. Burial of the dead is by hand in many areas and a priority for the community to prevent disease outbreak.

## One Month After the Event

### Module Two—Questions for Participants to Discuss in Small Groups

1. If power was out for a month, how would you and your family cope?

2. What would life in your community look like?

3. What would your organization do without power for this length of time? Would people go to work? How would you pay them?

4. How would communications be maintained?

5. What kind of product and service exchange would take place between organizations?

6. How might you or your organization function under a Declaration of Emergency or martial law?

7. At the end of 30 days, what capabilities have you been able to restore?

8. How do you know what to expect in the future?

Notes

# Notes

Notes

# High-Altitude Electromagnetic Pulse
## Module Three

### Questions for Participants to Consider During This Module

1. Without power for three months, what would happen to you and your family? What actions would you take at this point?

2. What would life in your community look like? How would you work with your neighbors? Who would your neighbors at this point?

3. How would communications be maintained?

4. What information would you need at this point?

## Summary—Three Months

By heroic efforts, the military has managed to keep generators operational and fuel flowing to the nuclear power plants to avoid meltdowns of cores or cooling ponds. With military-protected and supported power teams gathered from all surrounding states, replacement/loaner transformers are being convoyed in and integrated. The entire nation has contributed most of their power substation restoration teams, and most of their available spare nontransformer hardware, as mandated by executive order. Power stations have been turned into command posts, and military logistics is being used to support and protect these teams as they begin to reintegrate some of the key areas. Inspection teams are in the field, designating some areas as off limits—other areas are seeing some limited re-occupancy.

The hydroelectric plant generators were the first to be restarted and some damaged transformers have been repaired or replaced. They now are able to provide power to begin black starting other portions of the electric power grid. Trouble-shooting and repair/replacement of damaged grid electronic monitoring and control systems is nearly complete to enable coordination and control of grid black-start activities. Electric power is operational in limited pockets of the nation. Complete restoration of the grid will take years due to the time necessary to manufacture, move, and install replacement high-voltage transformers for the electric transmission network. The nation has been forever changed.

Note: If the planning team would like to take this scenario further into the future, say for six or nine months or a year, they may want to use the same questions as were posed for month two. As the time without power lengthens, there are other issues that arise that the team may want to discover and discuss.

# Three Months After the Event

## Module Three—Questions for Participants to Discuss in Small Groups

1. Without power for three months, what would happen to you and your family? What actions would you take at this point?

2. What would life in your community look like? How would you work with your neighbors? Who would your neighbors at this point?

3. How would communications be maintained?

4. What information would you need at this point?

# Notes

# Notes

# Twelve Months After the Event

A year or longer outage could be caused by a large initial event or a series of multiple events over time. In addition, cascading effects could keep the power grid from full restoration.

## In the small groups discuss the following questions:

1. What is the new normal for you, your community, and your organization?

2. What happens based on experiences around the world when there is a long-term deprivation of resources?

# Notes

# Questions After Participating in the Exercise

1. What changes can you make at home and in your organization to be more prepared?

2. What interactions do you want to have within your community including other organizations to be more prepared?

3. What if one community were able to make its own power:
   a. Would it become the emergency operations center for the state?
   b. How could it be protected?
   c. Should communities try to make their own power?

# Notes

# Conclusion and Surveys

In response to this exercise, participants may want to know what they can do to make a difference. Appendix A contains some suggestions for action that participants could tailor to their circumstances.

The InfraGard EMP SIG creators of this exercise would appreciate feedback from both leaders and participants, so that the exercise material can be improved. Vulnerabilities and threats may change over time as may mitigation and preparation efforts. What changes in the industry and the environment should be included in future versions of the exercise?

Please complete these two surveys:

1. For leaders: **https://www.surveymonkey.com/s/YNYZYH5**. This survey will help to continually improve the exercise scenarios and documentation.

2. For participants: **https://www.surveymonkey.com/s/Y5HK9MH**. This is a sample that you could use for your participants. If you create your own survey to help improve the exercise, please send the results to **igempsig@infragardmembers.org**.

In addition, if you have questions or comments that you would like to send directly to us, please use the following email address: **igempsig@infragardmembers.org**.

# Appendix A: Possible Actions

Time and distance have historically buffered the North American continent from major threats. Today, however, cyber, EMP, and space weather threats can arrive in the homeland at the speed of light. This fast speed means that whatever countermeasures and preparations are in place at the onset of the event are those that will be available. Today, protection is far more important for reducing casualties than response and will enhance recovery. (This was supported by the Sage economic impact assessment that shows that protecting even 10% of the most critical infrastructure could avoid up to 85% of the economic loss in a medium impact event.)

Three categories of protection need to be considered:

1.  The first is to harden the existing centralized infrastructure. Each person has some ability to influence this, either by urging elected representatives to legislate the necessary changes, or to influence electric utility and other suppliers of infrastructure services. [*Note:* Advance preparation by presenters could include summarizing the names and contact information of utility executives and legislators. Presentations by early adopters of advanced preparations and best practices may be highlighted. *InfraGard members must not distribute or recommend specific recommendations for lobbying elected representatives.*]

2.  Even if the regional electric power grids were far more hardened for key threats such as these, the nation would still be vulnerable to failure of an overly centralized infrastructure. The InfraGard EMP SIG believes that the best method for assuring infrastructure operation is to reduce reliance on centralized infrastructures. Increasing the share of power that is generated by distributed sources, for example, may be one of the most significant ways to reduce vulnerability (e.g., microgrids). Utilities are moving toward adoption and integration of distributed energy systems that could be operate in island mode when larger portions of the grid fails. Individual communities, critical infrastructure facilities, and businesses can do the same.

3.  Decentralization includes becoming more self-reliant. For example, few people realize that everyone is responsible for obtaining clean water, food, electric power, and other resources every day. For obvious economic reasons, and because few people possess the skills and resources to provide for themselves, these tasks are largely outsourced. Consider that in order to survive:

    *   each person needs at least one gallon per day of drinking water;
    *   nobody performs well after three days without food;
    *   shelter is necessary to prevent hypothermia, stay dry, etc.;
    *   medications are essential for some people's survival;
    *   assuming you have organized these resources, what is your plan for when the unprepared decide they want you to share?

4.  Work ahead with your neighbors to be prepared:

    *   get to know your neighbors and discuss working together in times of emergencies or disasters;
    *   determine what special needs you or they have;
    *   determine what special skills you or they have;
    *   determine what special emergency communications equipment you or they possess;
    *   become acquainted and work with emergency management professionals at every level (town, country, state, and federal);
    *   work with your professional organizations that assist in business continuity and contingency planning;

- support youth programs within schools and others such as scouting organizations to develop preparation skills;
- discover and work with faith-based organizations to work through disaster preparation issues and opportunities;
- discover and work with local fire halls and veterans organizations to increase volunteer participation.

Progress in all of these actions will take a long time and will need to be maintained and continually improved. This makes it essential for each person to take actions that are both affordable and enjoyable in order to remain motivated and effective.

# Appendix B: Alphabetical List of Contributors

**Dr. George Baker III** has been involved in electromagnetic protection research, management, and education since 1973. He is professor emeritus of applied science at James Madison University, where he served as Technical Director of the university's Institute for Infrastructure and Information Assurance (IIIA). He is involved in research grants from industry and government in the areas of critical infrastructure assurance, high power electromagnetics, and nuclear/directed energy weapon effects including vulnerability assessments of major military and emergency communication facilities. He is the former director of the Defense Threat Reduction Agency's Springfield Research Facility (SRF), where he was responsible for multi-hazard vulnerability assessment and protection of critical U.S. and NATO defense facilities and mobile systems. At SRF, he was instrumental in organizing the initial Force Protection Program for the Joint Chiefs of Staff. As Chief of the Defense Nuclear Agency's Innovative Concept Division, Dr. Baker was involved in early research on electromagnetic guns for the U.S. Army and Navy and managed the joint U.S.–Russian TOPAZ space nuclear power program. As the Electromagnetic Protection Team leader within the Defense Nuclear Agency's Radiation Effects Directorate, he led the development of EMP protection standards including MIL-STD-188-125, MIL-STD-2169B, and MIL-HDBK-423. He also led DOD efforts to protect ICBM/BMD systems and their associated communications, assess EMP effects on the national electric power grid, and develop and demonstrate RF weapon technology. He served on the Congressional EMP Commission senior staff and is presently a member of the National Defense Industrial Association (NDIA) Homeland Security Executive Board and the Infrastructure Security Partnership (TISP). He is a founding member of the Virginia Alliance for Secure Computing and Networking (VA SCAN) and the Directed Energy Professional Society. Dr. Baker is a fellow of the Summa Electromagnetics Foundation and senior member of the IEEE.

**Terry Donat, MD,** is a dual-board certified Facial Plastic and Reconstructive Surgeon/OHNS and Medical Investigator practicing in Northern Illinois and greater metropolitan Chicago for the past 15 years. Dr. Donat trained as a Biochemist and received his Medical Degree in Philadelphia in 1991. Terry is a written-exam reviewer and oral board examiner of U.S. and Canadian surgeons for the American Board of Facial Plastic and Reconstructive Surgery. He has extensive experience in facial reconstruction and managing blunt/penetrating head and neck trauma. Terry is the first physician certified as an Illinois Professional Emergency Manager. He is trained in acute Radiation Emergency Medicine, the Medical Management of Chemical and Biological Casualties and as a past National Disaster Life Support Instructor. He is appointed IEMA RACES Regional Radio Officer on the State Team for IEMA Region 2. Terry is the first physician to complete the graduate program in Veterinary Homeland Security from Purdue University—National Biosecurity Resource Center. He is a lifetime member of the Special Operations Medial Association (SOMA) and is currently focused on novel means for mitigating heat stress in austere environments and extending force protection while wearing PPE. Terry serves as the Acting Sector Chief—Health and Public Health for InfraGard Chicago; serves aside James Terbush, MD, MPH, as co-chair of the Healthcare Industry Advisory Group of the InfraGard EMP Special Interest Group; and is a member of the Healthcare and Public Health Sector Coordinating Council—DHS/DHHS. He is keenly interested in Biosecurity and Medical Intelligence and assessing the threats, risks, and vulnerabilities of evolving dual-use technologies.

**Dave Hunt** has led the development of several electrical grid exercises, including for the Congressional EMP Caucus, the National Defense University, Maryland Emergency Management Agency/Johns Hopkins Applied Physics Lab, the Electric Infrastructure Security Council, the Commonwealth of Pennsylvania, and the State of Tennessee. Mr. Hunt has 27 years of experience in all phases of emergency management, including the development of terrorism response plans for emergency response and national preparedness guidance. His response experience includes law enforcement, terrorism response, explosives investigation, fire and arson investigation, hazardous materials response, and emergency medical response. He has conducted planning projects for FEMA and its predecessor agencies with all 56 states and U.S. territories. Over the last 15 years, his work has addressed all phases of disaster response, and he has led multiple programs to develop national guidance for strategic response planning, as well as many of FEMA's Comprehensive Preparedness Guides (CPGs). He has led the creation of intelligence fusion guidance, as well as mitigation, response, and recovery planning on behalf of FEMA.

He has managed Improvised Nuclear Device planning for FEMA's National Preparedness Directorate, leading a large team of employees and dozens of consultants and subcontractors, including Lawrence Livermore National Laboratory and Virginia Tech's Center for Technology Security and Policy. He has piloted several IND projects across the nation, including pre-event public education, evacuation planning, and calculating the impacts to a community from detonation of an improvised nuclear device. He led a series of radiological exercises and supported nuclear response planning for the New York City region. He has led the development of Pennsylvania and Tennessee's Energy Assurance planning and associated exercises. Mr. Hunt has conducted training for the FBI, the Justice Department, multiple branches of the military, DHS, FEMA, and numerous states in support of homeland security preparedness efforts.

**William Kaewert** is founder of two power protection companies and has over 30 years of experience applying technology-based solutions that assure continuity of electrical power to critical applications. He is currently president and chief technology officer of Colorado-based Stored Energy Systems LLC (SENS), an industry leading supplier of nonstop DC power systems essential to electric power generation and other critical infrastructures. The company also produces COTS-based power converters used in EMP-hardened military systems including ground power for Minuteman III ICBM and THAAD ballistic missile interceptor. He received his AB in history from Dartmouth College and MBA from Boston University. He serves on the board of directors of the Electrical Generation Systems Association (EGSA) and on the management team of the Federal Bureau of Investigation's InfraGard Electromagnetic Pulse Special Interest Group (EMP SIG).

**Mary D. Lasky** is a Certified Business Continuity Professional (CBCP). Mary is the Program Manager for Business Continuity Planning for the Johns Hopkins University Applied Physics Laboratory (JHU/APL), and also coordinated the APL Incident Command System Team. Mary is President of the Community Emergency Response Network Inc. (CERN) in Howard County, Maryland. Mary is the immediate Past President of the Central Maryland Chapter of the Association of Contingency Planners (ACP). She is a member of InfraGard and on the executive committee for the InfraGard EMP SIG. She is a member of the FEMA Nuclear-Radiation Communications Working Group. Mary has held a variety of supervisory positions in Information Technology and in business services. For many years, she has been on the adjunct faculty of the Johns Hopkins University Whiting School of Engineering, teaching in the graduate degree program in Technical Management. Mary is the President of the Board of Directors of Grassroots Crisis Intervention Center in Howard County, MD. She is on the Finance Committee for Leadership Howard County and is co-chair of the Steering Committee for the Leadership Premier Program. Mary's consulting work has included helping nonprofit organizations create and implement their business continuity plans. She is the president of MDL Strategic Solutions LLC.

**Cedric Leighton** (USAF, Ret) is an internationally known strategic and cyberrisk expert who has appeared on CNN, FOX News Channel, Fox Business Network, Bloomberg TV, CNBC, PBS, BBC Al Jazeera, CCTV (China), C-Span, MSNBC, and other international and domestic broadcast outlets. He has written articles on cyberstrategy, national security, and management topics for publications such as *AOL Defense, Business Excellence Magazine,* TheHill.com, *MWorld, Leadership Excellence Magazine,* theStreet.com, CNBC.com, and the *Huffington Post.*

As the Founder and Chairman of Cedric Leighton International Strategies, he has facilitated the adoption of critical cybersecurity projects for domestic and international clients. He served 26 years as an Intelligence Officer in the U.S. Air Force, where he received numerous awards, including the Defense Superior Service Medal and the Bronze Star. Colonel Leighton holds a Bachelor's Degree in History (magna cum laude) and German Area

Studies from Cornell University, and a Master's Degree in International Studies from Angelo State University. He is National Journal "National Security Insider," a contributor to "America's Morning News," and a member of the Global Association of Risk Professionals (GARP).

**Chuck Manto** is the CEO of the Maryland-based company Instant Access Networks LLC (IAN). Mr. Manto won six patents in information, telecommunications and EMP shielding with others pending on microgrids and EMP shielding. He founded the InfraGard National EMP special interest group (SIG) and serves as its volunteer national manager. His prior management experience includes managing operations for a competitive local exchange carrier. Mr. Manto's education includes a BA and MA from the University of Illinois at Urbana/Champaign. He is a Senior Member of the IEEE.

**Thomas Popik** is Chairman of the Foundation for Resilient Societies, a nonprofit group dedicated to the protection of critical infrastructure against infrequently occurring natural and manmade disasters. He is principal author of a Petition for Rulemaking submitted to the Nuclear Regulatory Commission that would require backup power sources for spent fuel pools at nuclear power plants. Previously, as a U.S. Air Force officer, Mr. Popik investigated unattended power systems for remote military installations. Mr. Popik graduated from MIT with a B.S. in mechanical engineering and from Harvard Business School with an M.B.A. Mr. Popik can be reached at thomasp@resilientsocieties.org

**Mr. Dana Reynolds** is a 15-year veteran of the Colorado Department of Public Safety. He began his career as a State Trooper with the Colorado State Patrol. He held assignments as a policy and research specialist, accreditation manager, field supervisor, internal affairs investigator and polygraphist, Sergeant, Captain of the Homeland Security Section, and Director of the Colorado Information Analysis Center (CIAC), the state fusion center. Mr. Reynolds then accepted a position as Director of the Office of Preparedness within DHSEM. He was most recently promoted to Deputy Director. In his current role as Deputy Director of the Colorado Division of Homeland Security & Emergency Management, Mr. Reynolds has direct oversight over several vital programs, including geospatial analysis and imagery; state infrastructure protection; cybersecurity; state preparedness; interoperable communications; grant and finance; policy, strategic planning, and rulemaking; and strategic communications. Mr. Reynolds holds a bachelor's degree in Public Administration and a master's degree in Criminal Justice. He is also a graduate of the Harvard Kennedy School's Program on Crisis Leadership, an executive education program, and is currently enrolled in the Center for Homeland Defense & Security's Executive Leadership Program at the Naval Postgraduate School in Monterey, CA.

**Bob Rutledge** is the Lead of the Space Weather Forecast Office at NOAA's Space Weather Prediction Center (SWPC) in Boulder Colorado. SWPC is the United States' official source for civilian space weather watches, warnings, and alerts. Prior to joining SWPC, Bob worked at NASA's Johnson Space Center as the International Space Station (ISS) Radiation System Manager, responsible for oversight of the development and sustaining engineering of NASA's operational radiation measurement hardware onboard ISS. Bob began his career at NASA with the Space Radiation Analysis Group with responsibilities spanning planning, modeling, measurement, and operational management of astronaut radiation exposures. Bob received his degree in Electrical Engineering from Iowa State University.

# Appendix C: Reference Material

*Much of this list is accessible, along with additional resources, through links to InfraGard National EMP SIG through the secure InfraGard portal at www.InfraGard.org and at http://www.ipsonet.org/conferences/the-dupont-summit*

Additional material can be found in the annual conference proceedings of the EMP SIG sessions of the Dupont Summit through the series "High Impact Threats" available in print and Kindle editions through Westphalia Press and Amazon.

## Presentations and Introduction to the Latest Science on EMP, Its Impacts, and Related Policies

Slide presentations by presenters to the December 4, 2014 tabletop exercise and the December 5, 2014 InfraGard National EMP Sessions of the Dupont Summit are available at this DHS website by signing in as a guest: **https://share.dhs.gov/infragard-empsig-1204-052014/**

**Federal Energy Regulatory Commission.** *Electromagnetic Pulses: Effects on the U.S. Power Grid.* January 2010. The Federal Energy Regulatory Commission (FERC) released a study consisting of a series of comprehensive technical reports produced by Oak Ridge National Laboratory (for FERC) in joint sponsorship with the Department of Energy (DOE) and the Department of Homeland Security (DHS). It was issued to provide the basis for a technical understanding of how EMP threats affect the power grid. The Executive Summary and underlying reports are available here:

- Executive summary: **http://empactamerica.org/ferc_Executive_Summary.pdf**
- *Geomagnetic Storms and Their Impacts on the U.S. Power Grid* (Meta-R-319). John Kappenman. Metatech Corporation, January 2010. **http://www.ornl.gov/sci/ees/etsd/pes/pubs/ferc_Meta-R-319.pdf**
- *The Early Time (E1) High-Altitude Electromagnetic Pulse (HEMP) and Its Impact on the U.S. Power Grid* (Meta-R-320). Edward Savage, James Gilbert, William Radasky. Metatech Corporation, January 2010. **http://www.ornl.gov/sci/ees/etsd/pes/pubs/ferc_Meta-R-320.pdf**
- *The Late-Time (E3) High-Altitude Electromagnetic Pulse (HEMP) and Its Impact on the U.S. Power Grid* (Meta-R-321). James Gilbert, John Kappenman, William Radasky and Edward Savage. Metatech Corporation, January 2010. **http://www.ornl.gov/sci/ees/etsd/pes/pubs/ferc_Meta-R-321.pdf**
- *Low-Frequency Protection Concepts for the Electric Power Grid: Geomagnetically Induced Current (GIC) and E3 HEMP Mitigation* (Meta-R-322). John Kappenman. Metatech Corporation, January 2010. **http://www.ornl.gov/sci/ees/etsd/pes/pubs/ferc_Meta-R-322.pdf**
- *Intentional Electromagnetic Interference (IEMI) and Its Impact on the U.S. Power Grid* (Meta-R-323). William Radasky, Edward Savage. Metatech Corporation, January 2010. **http://www.ornl.gov/sci/ees/etsd/pes/pubs/ferc_Meta-R-323.pdf**
- *High-Frequency Protection Concepts for the Electric Power Grid* (Meta-R-324). William Radasky, Edward Savage. Metatech Corporation, January 2010. **http://www.ornl.gov/sci/ees/etsd/pes/pubs/ferc_Meta-R-324.pdf**

**North American Electric Reliability Corporation.** *High-Impact, Low-Frequency Event Risk to the North American Bulk Power System.* June 2010. It is a Jointly Commissioned Summary Report of the North American Electric Reliability Corporation and the U.S. Department of Energy's November 2009 Workshop. Attendants included representatives from various federal agencies and the North American electric industry's major sectors. The workshop discussed the risks of cyber or physical coordinated attack, pandemic, and geomagnetic disturbance/EMP. The discussion focused on approaches to measure and monitor HILF risks, potential mitigation steps, and formulating an effective public/private partnership to more effectively address these issues. The summary of the report can be found here: **http://www.nerc.com/files/HILF.pdf**

## Additional Resources and References (arranged by source)

Academy of Science Report: *Severe Space Weather Events, Understanding Societal and Economic Impacts*—Workshop Report, May 2008. A public workshop with representatives of industry, government, and academia was held to discuss socioeconomic impacts of extreme space weather events and to address the questions of space weather risk assessment and management. The discussion included direct and collateral effects of severe space weather events, the current state of the space weather services infrastructure in the United States, the needs of users of space weather data and services, and the ramifications of future technological developments for contemporary society's vulnerability to space weather. The report of this workshop can be read here:
**http://books.nap.edu/openbook.php?record_id=12507&page=1**

Baker G. 2014. *Electromagnetic Pulses—Six Common Misconceptions.* Domestic Preparedness.
**http://www.domesticpreparedness.com/Commentary/Viewpoint/Electromagnetic_Pulses_-_Six_Common_Misconceptions/**

Defense Science Board. April 2005. *Report of the Defense Science Board Task Force on Nuclear Weapon Effects—Tests, Evaluation, and Simulation.* U.S. Department of Defense. The Task Force identified nuclear weapon effects test and simulation needs of Department of Defense in current and emerging threat environments and produced a roadmap of capabilities the guide the development of future simulators and simulation technology. The report can be found at: **http://www.survive-emp.com/fileadmin/White-Papers/EMP-Resources/DoD-Science-Task-Force-apr05.pdf**

Electromagnetic Pulse Commission. 2008. *Report of the Commission to Assess the Threat to the United States from Electromagnetic Pulse (EMP) Attack.* Critical National Infrastructures. This report presents the results of the EMP Commission's assessment of the effects of a high-altitude electromagnetic pulse (EMP) attack on the U.S. critical national infrastructures and provides recommendations for their mitigation.
**http://www.empcommission.org/docs/A2473-EMP_Commission-7MB.pdf**

FEMA. February 2010. *The Workshop on Managing Critical Disasters in the Transatlantic Domain—The Case of a Geomagnetic Storm.* This workshop brought together partners from the European Union, European Commission, Swedish government, U.S. National Weather Service (NWS), and the U.S. National Oceanic and Atmospheric Agency (NOAA). Participants discussed readiness and planning recommendations for geomagnetic storms. The summary of this workshop can be downloaded from the FEMA Library:
**http://www.fema.gov/library/viewRecord.do?id=4270**

*Interdependence of the Electricity Generation System and the Natural Gas System and Implications for Energy Security* (Judson N.: MIT Lincoln Laboratory, 2013). **https://www.serdp-estcp.org/content/download/19069/.../TR-1173.pdf**

Manto C. 2014. *Solar Storm Near Miss and Threats to Lifeline Infrastructure.* Domestic Preparedness.
**http://www.domesticpreparedness.com/Infrastructure/Cyber_%26_IT/Solar_Storm_Near_Miss_%26_Threats_to_Lifeline_Infrastructure/**

Metatech Corporation. 2010. "The Early-Time (E1) High-Altitude Electromagnetic Pulse (HEMP) and Its Impact on the U.S. Power Grid." Prepared for Oak Ridge National Laboratory. **http://www.ferc.gov/industries/electric/indus-act/reliability/cybersecurity/ferc_meta-r-320.pdf**

Parformak P. 2014. *Physical Security of the U.S. Power Grid: High-Voltage Transformer Substations*. Congressional Research Service. **http://fas.org/sgp/crs/homesec/R43604.pdf**

National Science and Technology Council; "National Space Weather Strategy", 2015: **http://www.dhs.gov/national-space-weather-strategy**

*Severe Space Weather Events—Understanding Societal and Economic Impacts*. NRC-NAP, 2008. **http://books.nap.edu/catalog/12507/severe-space-weather-eventsunderstanding-societal-and-economic-impacts-a-workshop**

Smith R. 2014. "U.S. Risks National Blackout From Small-Scale Attack: Federal Analysis Says Sabotage of Nine Key Substations Is Sufficient for Broad Outage." *Wall Street Journal*. **http://online.wsj.com/articles/SB10001424052702304020104579433670284061220**

*Solar and Space Physics: A Science for a Technological Society*. NRC-NAP, 2013. **http://www.nap.edu/catalog/13060/solar-and-space-physics-a-science-for-a-technological-society**

*Solar Storm Risk to the North American Electrical Grid*. Lloyds of London, 2013. **http://www.lloyds.com/~/media/Lloyds/Reports/Emerging%20Risk%20Reports/Solar%20Storm%20Risk%20to%20the%20North%20American%20Electric%20Grid.pdf**

Space Weather Prediction Center web site, National Weather Service, NOAA. **http://origin-www.swpc.noaa.gov/**

*Terrorism and the Electrical Power Delivery System*. NRC-NAP, 2012. **http://www.nap.edu/catalog/12050/terrorism-and-the-electric-power-delivery-system**

*The EMP Threat: Examining the Consequences*. Subcommittee on Cybersecurity, Infrastructure Protection and Security Technologies of the Committee on Homeland Security, House of Representatives, 112th Congress, 2012. **http://www.gpo.gov/fdsys/pkg/CHRG-112hhrg80856/html/CHRG-112hhrg80856.htm**

*The U.S. Government Thinks China Could Take Down the Power Grid*. **http://www.cnn.com/2014/11/20/politics/nsa-china-power-grid/index.html?hpt=hp_t2**

Wilson Clay. "High-Altitude Electromagnetic Pulse (HEMP) and High Power Microwave (HPM) Devices: Threat Assessments." *Congressional Research Service Report for Congress,* July 2008. **http://www.survive-emp.com/fileadmin/White-Papers/EMP-Resources/crs-RL32544-emp-jul08.pdf**

## Electromagnetic Pulses From Nuclear Detonation (arranged by author)

Armed Forces Special Weapons Project. July 1958. "Operation of Balloon Carrier for Very-High-Altitude Nuclear Detonation." *Report* ITR-1655 (Re-designated ADA322231).

Bainbridge, K.T. May 1976. "Trinity." *Report LA-6300-H,* Los Alamos Scientific Laboratory, 53.

Baum, Carl E. 1992. "From the Electromagnetic Pulse to High-Power Electromagnetics." *Proceedings of the IEEE* 80 (6): 789-817.

Baum, Carl E. 2007. "Reminiscences of High-Power Electromagnetics." *IEEE Transactions on Electromagnetic Compatibility* 49 (2): 211-218.

Broad, William. 1981. "Nuclear Pulse (I): Awakening to the Chaos Factor." *Science*, Vol. 212, May, pp. 1009-1012.

Broad, William. 1981. "Nuclear Pulse (II): Ensuring Delivery of the Doomsday Signal." *Science*, Vol.212, June pp.1116-1120.

Broad, William. 1981. "Nuclear Pulse (III): Playing a Wild Card." *Science*, Vol. 212, June pp. 1248-1251. **http://www.sciencemag.org/content/212/4498/1009.citation**

Defense Atomic Support Agency. September 1959. Operation Hardtack, Preliminary Report, ITR-1660(SAN) (Re-designated ADA369152). (Also see the Chapter 10 pages on Yucca EMP measurements.)

Defense Nuclear Agency. 1958. "Operation Hardtack High-Altitude Test Film, Includes Films of the Hardtack-Yucca Balloon-Launched 1.7 kiloton Nuclear Test, as Well as the Redstone Missile Launched Hardtack-Teak and Hardtack-Orange 3.8 megaton High-Altitude Tests." *Report* DNA 6038F, Operation Hardtack, 1958.

Department of Defense. 1994. "High-Altitude Electromagnetic Pulse (HEMP) Protection for Ground-Based C4I Facilities Performing Critical, Time-Urgent Missions." Military Standard 188-125A. **https://www.google.com/#q=mil+spec+188.125**

Dyal, P. 1965. "Operation Dominic. Fish Bowl Series. Debris Expansion Experiment." *Report* ADA995428, Air Force Weapons Laboratory, 15.

Federal Emergency Management Agency. 1991a. "The Theoretical Basis for EMP Protection." *Electromagnetic Pulse Protection Guidance*, Volume 1 (CPG 2-17)' Washington, DC, (See **training.fema.gov**)

Federal Emergency Management Agency. 1991. "EMP Protection Applications." *Electromagnetic Pulse Protection Guidance*, Volume 2 (CPG 2-17) Washington, DC, (See **training.fema.gov**)

Federal Emergency Management Agency. 1991. "EMP Protection Installations." *Electromagnetic Pulse Protection Guidance*, Volume 3 (CPG 2-17) Washington, DC, (See **training.fema.gov**)

Gilinsky, Victor. 1964. *The Kompaneets Model for Radio Emission from a Nuclear Explosion.* The Rand Corporation, Marina del Rey.

Kompaneets, Aleksandr S. 1958. "Radio Emission from an Atomic Explosion." *Journal of Experimental and Theoretical Physics,* Issue 8, Volume 1076 from Zhournal Eksperiment. i. Teori. Fiz #35, Tbilisi, USSR. (Translated from Russian-Language Publication).

Longmire, Conrad L. 2004. "Fifty Odd Years of EMP." NBC Report, U.S. Army Nuclear and Chemical Agency, 47-51.

Longmire, Conrad L. "Theoretical Notes—Note 353—March 1985—EMP on Honolulu from the Starfish Event." Mission Research Corporation.

Miller, C. 2005. "Electromagnetic Pulse Threats in 2010." *Research Paper.* Air War College Maxwell Air Force Base, Center for Strategy and Technology.

Rabinowitz, Mario. 1987. "Effect of the Fast Nuclear Electromagnetic Pulse on the Electric Power Grid Nationwide: A Different View." *IEEE Transactions on Power Delivery* PWRD-2: 1199-1222.

Younger, Stephen, et al. 1996. "Scientific Collaborations Between Los Alamos and Arzamas-16 Using Explosive-Driven Flux Compression Generators." *Los Alamos Science* (Vol. 24): pp. 48-71.

U.S. Central Intelligence Agency. National Intelligence Estimate. Number 11-2A-63, The Soviet Atomic Energy Program, 44.

Vittitoe, Charles N. 1989. *Did High-Altitude EMP Cause the Hawaiian Streetlight Incident?* Sandia National Laboratories, New Mexico.

## Electromagnetic Effects from Space Weather

Baker, Daniel. 2014. "New Twists in Earth's Radiation Belts." *American Scientist* 102—Rings of high-energy particles encircling our planet change more than researchers realized. Those variations could amplify damage from solar storms. **http://www.americanscientist.org/issues/feature/2014/5/new-twists-in-earths-radiation-belts**

Brian, C., et al. 2011. "Revisiting the Carrington Event: Updated Modeling of the Atmospheric Effects." *Journal of Geophysical Research—Atmospheres.* **http://arxiv.org/abs/arXiv:1111.5590**

Dobbins, R., C.J. Schrijver, W. Murtagh, and M. Petrinic. 2014. "Assessing the Impact of Space Weather on the Electric Power Grid Based on Insurance Claims for Industrial Electrical Equipment." *Space Weather Journal.* Authors review 10 years of insurance claims linking damage to regular space weather effects. **http://onlinelibrary.wiley.com/enhanced/doi/10.1002/2014SW001066/**

Kemp, John; Reuters. February 18, 2014. "U.S. Orders Power Grid to Prepare for Solar Storms." **http://www.reuters.com/article/2014/02/18/electricity-solar-storms-idUSL6N0LN3HU20140218**

NASA. July 2012. "Solar Storm Near Miss." 2014. **http://science.nasa.gov/science-news/science-at-nasa/2014/23jul_superstorm/**

Ryan, Tracy. 2014. "Here Comes the Sun Storm." *Wall Street Journal.* **http://online.wsj.com/news/articles/SB10001424052702303505504577404360076098508**

Washington Post Editorial Board on Solar Storms. August 9, 2014. **http://www.washingtonpost.com/opinions/extreme-space-weather-threatens-to-leave-the-us-in-the-dark/2014/08/09/22782cd4-1c26-11e4-82f9-2cd6fa8da5c4_story.html**

## FERC Rulings and Publications

Rule 779 on Geomagnetic Disturbances, See 143 FERC 61,147, United State of America, Federal Energy Regulatory Commission, 18 CFR Part 40, [Docket No. RM12-22-000; Order No. 779], Reliability Standards for Geomagnetic Disturbances (Issued May 16, 2013). Updated standards can also be found at the FERC website: **http://www.ferc.gov/whats-new/comm-meet/2013/051613/E-5.pdf**

FERC Remand of NERC Cyber Regulations. **http://www.ferc.gov/whats-new/comm-meet/2013/032113/E-11.pdf**

142 FERC 61,204, United States of America, Federal Energy Regulatory Commission, Docket No. RD12-5-000 FERC, Executive Summary, Effects of EMP on Electric Power Grid. **http://www.ferc.gov/industries/electric/indus-act/reliability/cybersecurity/ferc-executive_summary.pdf**

FERC, LaFleur, Cheryl A., Acting Chairman before Committee on Energy and Commerce, Subcommittee on Energy and Power, U.S. House of Representatives, Hearing on the Role of FERC in a Changing Energy Landscape, December 5, 2013. **http://ferc.gov/CalendarFiles/20131205094201-LaFleur-12-05-2013.pdf**

"Joint Meeting of the Nuclear Regulatory Commission and the Federal Energy Regulatory Commission." AD06-6-000, Friday, June 15, 2012, 9:30–11:30 a.m. (covers discussion between FERC, NRC and NERC on solar storm effects on the grid. **http://www.nrc.gov/reading-rm/doc-collections/commission/tr/2012/20120615.pdf**

Nuclear Regulatory Commission. "Long-Term Cooling and Unattended Water Makeup of Spent Fuel Pools." NRC-2011-0069 (Phased rule making in response to Petition from Tom Popik on GMD impacts on ability to cool spent fuel rods.) **http://www.regulations.gov/#!documentDetail;D=NRC-2011-0069-0109**

Station Blackout Mitigation Strategies [NRC-2011-0299]. **https://www.federalregister.gov/regulations/3150-AJ08/station-blackout-mitigation-strategies-nrc-2011-0299**

The NRC published an Advance Notice of Proposed Rulemaking (ANPR) on March 20, 2012 (77 FR 16175), to seek public comments on potential changes to the Commission's regulations that address a condition known as station blackout (SBO). SBO involves the loss of all on-site and off-site alternating current (ac) power at a nuclear power plant. A central objective of this rulemaking would be to make generically applicable requirements previously imposed on licensees by EA-12-049 "Order Modifying Licenses with Regard to Requirements for Mitigating Strategies for Beyond-Design-Basis External Events," while ensuring that the new requirements are properly integrated with the existing SBO requirements in 10 CFR 50.63. This regulatory action is one of the near-term actions based on lessons-learned stemming from the March 2011, Fukushima Dai-ichi event in Japan. Includes references to GMD impacts.
**https://www.federalregister.gov/articles/2012/03/20/2012-6665/station-blackout#h-15**

## Additional Reference Material (arranged by appearance in Maine PUC docket below)

*Bibliography of Comments Filed with the Maine Public Utility Commission Docket 2013-00415 for LD 131 in 2013 (order of items below roughly replicates that within the public record).*

The Sage Policy Group. 2007. "Initial Economic Assessment of Electromagnetic Pulse (EMP) Impact upon the Baltimore–Washington–Richmond Region." *Economic Impact Report,* September 10.

Ambassador James Woolsey, R. 2013. "Maine Should Protect its Electric Grid from Electromagnetic Pulse (EMP)." Letter to State of Maine.

James Woolsey, R. 2013. "Testimony Before the House Committee on Energy and Commerce." Testimony, May 21.

Electric Infrastructure Security Council. 2013. "Inquiry into Measures to Mitigate the Effects of Geomagnetic Disturbances and Electromagnetic Pulse on the Transmission System in Maine." Comments for MPUC and Attachments, October 4.

Advanced Fusion Systems LLC. 2013. "Advanced Fusion Systems LLC Hardware and Test Capabilities." Statement of Capabilities, October 4.

Resilient Societies Press Release.June_16, 2014. Transmitted-2.
**http://www.resilientsocieties.org/pressreleases.html**

Foundation for Resilient Societies. 2013. "Comments of The Foundation for Resilient Societies in Response to 14 Questions Propounded by the Public Utilities Commission of the State of Maine Together with Appendices." Comments to MPUC, October 4.

Foundation for Resilient Societies. 2013. "Response to NERC Request for Comments on Geomagnetic Disturbance Planning Application Guide." Appendix 1 of 4 to Comments to MPUC, August 9.

Foundation for Resilient Societies. 2012. "Comments of The Foundation for Resilient Societies Before the Federal Energy Regulatory Commission, Reliability Standards for Geomagnetic Disturbances." Appendix 2 of 4 to Comments to MPUC, Docket No. RM12-22-000, December 24.

Foundation for Resilient Societies. 2013. "Comments of The Foundation for Resilient Societies Before the Federal Energy Regulatory Commission, Reliability Standards for Geomagnetic Disturbances." Appendix 3 of 4 to Comments to MPUC, Docket No. RM12-22-000, April 1.

Foundation for Resilient Societies. 2013. "Comments of The Foundation for Resilient Societies Before the Federal Energy Regulatory Commission, Reliability Standards for Geomagnetic Disturbances." Appendix 4 of 4 to Comments to MPUC, Docket No. RM12-22-000, May 14.

Emprimus LLC. "Resolve, Directing the Public Utilities Commission To Examine Measures To Mitigate the Effects of Geomagnetic Disturbances and Electromagnetic Pulses on the State's Transmission System—Interim Report." Answers to Questions Arising from MPUC Interim Report, LD 131.

Task Force on National and Homeland Security. 2013. "Rebuttal to Public Utilities Commission Report That Recommends Doing Nothing to Protect the Maine Electric Grid from Electromagnetic Pulse (EMP) and Other Threats." Comments by Dr. Peter Vincent Pry, December 10.

EMP Coalition. "Maine State Legislature Warning Against 'improvements'." Letter, Maine PUC.

Cynthia Ayers. "Relating to Geomagnetic Disturbance (GMD) and Electromagnetic Pulse (EMP)." Comments to MPUC, Comments on Public Utilities Commission Report IAW Resolves 2013, Ch. 45.

Chuck Manto Provided Transcripts of Speaker Comments from DuPont Summit, InfraGard National EMP SIG Sessions at the December 6, 2013 Dupont Summit, Washington, DC, December 6, 2013.

Emprimus LLC. 2013. "Response to Draft Report on GMD/EMP Risk to Maine Power Grid." Comments to MPUC, December 18.

Foundation for Resilient Societies. 2013. "Recommendations of The Foundation for Resilient Societies to Strengthen the Final Report of the Maine Public Utilities Commission to the Maine State Legislature on Mitigation of Geomagnetic Disturbances and Electric Magnetic Pulse Risks to the Maine Electric Grid." Recommendations to Maine Legislature, December 18.

Center for Security Policy. 2013. Statement by Frank Gaffney for the MPUC, Statement by Frank J. Gaffney, Jr., December 18.

Center for Security Policy. Web Log Entry by Frank Gaffney on Insights Gained from the 2013 DuPont Summit of the InfraGard National Electromagnetic Pulse Special Interest Group (EMP SIG).

Federal Energy Regulatory Commission. 2013. "Re: Maine Public Utilities Commission Initiated Inquiry into

Measures to Mitigate the Effects of Geomagnetic Disturbances and Electromagnetic Pulse on the Transmission System in Maine." Letter to MPUC, Docket No. 2013-00415, December 23.

"Civilian EMP Rating System and Sample Methods to Protect Control Systems and Networks." See Document {365F6715-9070-4A28-8889-AEC2F57F0595} at **https://mpuc-cms.maine.gov/CQM.Custom.WebUI/MatterFiling/MatterFilingItem.aspx?FilingSeq=78979& CaseNumber=2013-00415**

Bibliography of Filings of the Maine Public Utilities Commission. Docket 2013-00415 on LD131 Concerning Electromagnetic Pulse and Geomagnetic Storms.

Maine Public Utilities Commission. 2013. "Inquiry into Measures to Mitigate the Effects of Geomagnetic Disturbances and Electromagnetic Pulse on the Transmission System in Maine." Notice of Inquiry, August 21.

Metatech Corp. 2013. "Role of Assessments for Geomagnetic Storm Protection." William Radasky Memo to Andrea Boland, September 4.

ISO New England Inc. 2013. "Inquiry into Measures to Mitigate the Effects of Geomagnetic Disturbances and Electromagnetic Pulse on the Transmission System in Maine." Letter to Maine PUC, Requested Information from ISO New England in Docket No. 2013-00415, October 4.

Emprimus LLC. "Emprimus LLC Answers to Maine PUC Inquiry Questions." Memo to Maine PUC.

Foundation for Resilient Societies. 2013. "Comments on Reliability Standards for Geomagnetic Disturbances." Filing Submitted to FERC, Docket No. RM12-22-000, April 1.

Aon Benfield. 2013. "Geomagnetic Storms." *Risk Management Report,* January.

National Defense University. 2011. "Summarizing Research on the Threat of EMP to the U.S. Electric Grid." Richard Andres Letter to Andrea Boland, November 16.

Maine House of Representatives. 2013. "Andrea Boland Letter to Maine PUC with Linked Attachments." Comments in Support of PUC Docket 2013-00415, October 4.

Storm Analysis Consultants. "John Kappenman Memo to Maine PUC with Embedded Attachments." Comments of John G. Kappenman, Storm Analysis Consultants.

Foundation for Resilient Societies. 2013. "Written Testimony Before Joint Standing Committee on Energy, Utilities and Technology." State of Maine Legislature Regarding Legislative Document 131, An Act to Secure the Safety of Electrical Power Transmission Lines, Written Testimony, March 5.

Task Force on National and Homeland Security. 2013. "Dr. Peter Vincent Pry to Andrea Boland, "Quick Fix" EMP Protection for the Maine Electric Grid." Memo, March 28. Andrea Boland. "Maine PUC Referencing Multiple Supporting Documents." Memo, Comments on Maine PUC Docket 2013-00415, EMP/GMD Mitigation Study: LD 131.

Maine State Legislature. "Body of LD 131 Plus 1 of 2 Groups of Written Testimony."

Maine State Legislature. "2 of 2 Groups of LD 131." Written Testimony.

Andrea Boland. 2013. "Introduce LD131, An Act to Secure the Safety of Electrical Power Transmission Lines in Maine." Text of Speech to Maine Energy, Utilities, and Technology Committee, February 19.

Central Maine Power. 2013. "Inquiry into Measures to Mitigate the Effects of Geomagnetic Disturbances and Electromagnetic Pulse on the Transmission System in Maine." Letter to Maine PUC with Attachments, Docket No. 2013-415, Maine Public Utilities Commission, October 4.

Foundation for Resilient Societies. 2013. Comments to Maine PUC, Supplemental & Reply Comments of the Foundation for Resilient Societies Submitted to the Public Utilities Commission of the State of Maine, October 15.

Bangor Hydro Electric Company. 2013. "Comments of Bangor Hydro Electric Company and Maine Public Service Company." Memo to Maine PUC, October 4.

Task Force on National and Homeland Security. 2013. "Cynthia Ayers Comments to Maine PUC." Maine PUC Questions on GMD/EMP Study, October 4.

Electric Infrastructure Security Council. 2013. The International E-Pro Report; International Electric Grid Protection, a Report Summarizing the Status of National Electric Grid Evaluation and Protection Against Electromagnetic Threats in 11 Countries, September.

Andrea Boland. 2013. "Comments of Representative Andrea Boland." Memo to Maine PUC, October 15.

State of Maine PUC Office of the Public Advocate. 2013. "Comments of the Office of the Public Advocate." Memo to Maine PUC, October 15.

Task Force on National and Homeland Security. Paper from Dr. Peter Vincent Pry to Maine PUC, "A North Korean Nuclear Pearl Harbor?"

Task Force on National and Homeland Security. Paper from Dr. Peter Vincent Pry to Maine PUC, "Maine's Battle to Save America."

Task Force on National and Homeland Security. Memo by Dr. Peter Vincent Pry, Dr. Fred Iklé, Professor Cynthia Ayers, Brig. Gen. Kenneth Chrosniak with foreword by Dr. William R. Graham, Civil-Military Preparedness for an Electromagnetic Pulse Catastrophe.

Maine House of Representatives. 2013. Andrea Boland Letter to Maine PUC in Response to the Public Advocate's Comments on Docket 2013-00415, October 16.

Maine Public Utilities Commission. 2013. "Relating to Geomagnetic Disturbance and Electromagnetic Pulse." Draft Report. Draft Report in Accordance with Resolves, Ch. 45, December 6.

Ambassador Henry Cooper. 2013. "Comments of Ambassador Henry F. Cooper." Memo to Maine PUC, December 16.

Andrea Boland. 2013. "Comments of Representative Andrea Boland in Response to Draft Report." Memo to Maine PUC, December 18.

Maine Public Utilities Commission. 2014. "Regarding Geomagnetic Disturbances (GMD) and Electromagnetic Pulse (EMP)." Report. Report to the Legislature Pursuant to Resolves 2013, Ch. 45, January 20.

Central Maine Power. Appendix 1 of 2 to MPUC January 14, 2014 Report, GMD-EMP Risk Analysis, January 20, 2014.

NPCC Inc. Appendix 2 of 2 to MPUC, January 14, 2014. Report, Document C-15, Procedures for Solar Magnetic Disturbances Which Affect Electric Power Systems, January 11, 2007.

*A Cyber Innovation Labs Venture*

# Max-Survivability Center

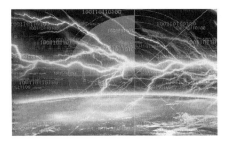

## Differentiation

- First commercially available Protected Platform from EMP/HEMP/IEMI and GMS events

- 360° tested, certified and shielded Tier 3+ Data Center rating

- Long-life survivability for mission critical apps & services

- SSAE16 SOC 2, PCI and HIPAA/HITECH compliance

- Value engineered at price points of traditional, legacy market offerings

- Real-dollar SLA offers $50K - $10M Uptime Guarantee

Enterprise-class data centers require sophistication in design, management, compliance, scalability and **Electromagnetic Pulse (EMP)** and **Geomagnetic Storm** protection to support a wide variety of client IT needs. The EMP Resilient Facility solution, coined by EMP GRID Services, leverages best-of-breed technologies to furnish enterprises with complete access, managed data capabilities and EMP hardening across the entire communications infrastructure.

## EMP Resilient Facility Delivers

- Tier 3+ availability, uncompromising security, reliability, and control supported by industry-leading SLAs.
- **Highest level of compliancy - PCI, HIPAA, HITRUST, SSAE 16 SOC 1 and SOC 2.**
- Six-tiered security perimeters, fully monitored 24x7, with live on-site support.
- Customizable and scalable solutions to ensure ongoing business support and reliability.
- Custom designed EMP/HEMP enclosures and shelters.
- 100% Private, single-tenant Cloud solution offers unparalleled performance and uptime (99.99%).

## EMP Resilient Facility Functions

- 360° tested, certified and shielded Tier 3+ Data Center platform.
- All MEP/FP (electrical, mechanical and fire protection) stand-alone with 2N redundancy.
- **$50K - $10M SLA guaranteed uptime performance per customer.**
- Primary emergency generation (diesel / natural gas / fuel cell) within shielded containment supported by **30 days of uninterrupted on-site fuel storage**.
- Complete telecom infrastructure contained and protected with mobile rollout deployment.
- Mirrored and fully managed EMP/Geomagnetic Storm protected recovery facilities.

## *Pre-Launch Sale*

Construction has begun on our 2,000 SF *Maximum Resilient*, **EMP and Geomagnetic Storm shielded data center**. Located in Mount Prospect, IL, approximately 25 miles outside of Chicago, this *Ultra Shielded* space will afford customers with **a next generation collocation and/or private cloud solution** for mission critical applications and services. Space will be ready for occupancy in early 2016.

*Contact us at (312) 593-6106 to learn more and ask about promotional pricing*

EMP GRID Services | www.empgridservices.com | P. 312.593.6106

Made in the USA
Middletown, DE
31 October 2015